Algorithms for Intelligent Systems

This book series publishes research on the analysis and development of algorithms for intelligent systems with their applications to various real world problems. It covers research related to autonomous agents, multi-agent systems, behavioral modeling, reinforcement learning, game theory, mechanism design, machine learning, meta-heuristic search, optimization, planning and scheduling, artificial neural networks, evolutionary computation, swarm intelligence, and other algorithms for intelligent systems.

The book series includes recent advancements, modification and applications of the artificial neural networks, evolutionary computation, swarm intelligence, artificial immune systems, fuzzy system, autonomous and multi agent systems, machine learning, and other intelligent systems related areas. The material will be beneficial for the graduate students, post-graduate students as well as the researchers who want a broader view of advances in algorithms for intelligent systems. The contents will also be useful to the researchers from other fields who have no knowledge of the power of intelligent systems, e.g. the researchers in the field of bioinformatics, biochemists, mechanical and chemical engineers, economists, musicians and medical practitioners.

The series publishes monographs, edited volumes, advanced textbooks and selected proceedings.

Indexed by zbMATH.

All books published in the series are submitted for consideration in Web of Science.

Manoj Sahni · José M. Merigó · Walayat Hussain
Editors

Novel Developments in Futuristic AI-based Technologies

Springer

Editors
Manoj Sahni
Pandit Deendayal Energy University
Gandhinagar, Gujarat, India

Walayat Hussain
Peter Faber Business School
Australian Catholic University
Sydney, NSW, Australia

José M. Merigó
School of Information, Systems
and Modelling
Faculty of Engineering and Information
Technology
University of Technology Sydney
Ultimo, NSW, Australia

ISSN 2524-7565 ISSN 2524-7573 (electronic)
Algorithms for Intelligent Systems
ISBN 978-981-99-3075-3 ISBN 978-981-99-3076-0 (eBook)
https://doi.org/10.1007/978-981-99-3076-0

This Springer imprint is published by the registered company Springer Nature Singapore Pte Ltd.
The registered company address is: 152 Beach Road, #21-01/04 Gateway East, Singapore 189721, Singapore

Preface

This book is a compilation of various topics related to mathematical tools for analysis and decision-making, applications of IoT and digital technology in real life, and applications of technology in medical and agricultural industries. The book covers a wide range of topics that are relevant in today's rapidly changing world, where technology is increasingly becoming a crucial part of our daily lives.

This book focuses on mathematical tools for analysis and decision-making, including topics such as fuzzy risk analysis assessment, in financial portfolio selection, forecasting exchange rates. These mathematical tools play a vital role in solving complex problems. Fuzzy risk analysis assessment, for example, is a mathematical tool that uses fuzzy logic to analyze and assess risks in decision-making problems. Fuzzy logic is a form of logic that allows for the representation of uncertain or imprecise information, making it particularly useful for risk analysis in complex systems. In financial portfolio selection, the multiple criteria hierarchical process is a mathematical tool used to select a portfolio of assets that meets specific criteria, such as risk and return. Forecasting exchange rates is another real-life problem that can be tackled using mathematical tools. One such tool is the hybridization of the ordered weighted averaging (OWA) operator with other methods, which has been shown to improve the accuracy of exchange rate forecasts in diffuse time series. Mathematical tools can also be used to study the properties and behavior of mathematical objects, such as the group of signed quadratic residues. This group has applications in cryptography, where it is used to construct cryptographic protocols that are resistant to certain types of attacks.

The book also delves into the applications of IoT and digital technology in real life, including topics such as cloud-assisted IoT infrastructure with anonymous key issuing, which provides a secure and efficient method of authentication for IoT devices. This approach enables IoT devices to communicate with each other securely, ensuring data privacy and reducing the risk of cyber-attacks. Another application of IoT and digital technology is providing assisted living for visually challenged persons. The Internet of Things (IoT)-based digital life system enables the visually challenged to live an independent life with improved quality of life. Information and Communication Technology (ICT) has also impacted tourist spending, with tourists increasingly relying on technology to plan and book their trips. Smartphones have become a ubiquitous device in modern society, and their use in retail-purchase buying intentions has been extensively studied. These technologies discussed in section two are transforming the way we live, work, and interact with our environment, and have the potential to revolutionize various industries.

The book also explores the applications of technology in medical and agricultural industries, including topics such as the development of automated breast cancer detection systems, which use machine learning algorithms to analyze mammography images and identify potential tumors. Another application of technology discussed in this book is the Proteomic analysis in the medical industry, which enables the classification and prediction of breast cancer survival rates. Similarly, one can find applications in other industries such as technology, which is being used to manage inventory during pandemics through passive RFID tags, which enable real-time tracking of inventory and reduce the risk of stock-outs. Blockchain is another technology discussed here, which is being used to enhance agriculture supply-chain management, enabling the tracking of products from farm to table and improving transparency and efficiency. These technologies are revolutionizing the way we diagnose and treat diseases, as well as improving the efficiency and sustainability of agricultural practices.

The contributors to this book are experts in their respective fields, and their diverse perspectives and experiences provide valuable insights into the latest trends and developments in the areas of mathematical tools for analysis and decision-making, applications of IoT and digital technology in real life, and applications of technology in medical and agricultural industries. These experts have conducted extensive research, applied their knowledge and skills in solving real-world problems, and shared their experiences and expertise in this book.

We believe that this book will serve as a valuable resource for researchers, students, and practitioners who are interested in these topics. The in-depth analysis, practical applications, and case studies presented in this book will enable readers to gain a comprehensive understanding of the latest developments and emerging trends in these fields.

For researchers, this book provides a wealth of information and insights that can serve as a basis for further research and exploration. Students will find this book an essential resource for understanding the underlying concepts, theories, and applications of mathematical tools for analysis and decision-making, IoT and digital technology, and technology in medical and agricultural industries. Practitioners, including professionals, managers, and executives, will benefit from the practical insights, real-world examples, and case studies presented in this book, which can help them to improve their decision-making and problem-solving skills.

Overall, we hope that this book will be a valuable addition to the existing literature in these fields and will inspire readers to further explore and contribute to the advancement of these important areas of research and practice.

Gandhinagar, India Dr. Manoj Sahni
Ultimo, Australia Dr. José M. Merigó
Sydney, Australia Dr. Walayat Hussain

Acknowledgements

We are incredibly thankful to the authors whose insightful research has been integral in creating this book. Their invaluable contributions have been absolutely essential and have enabled us to produce this book. We are profoundly grateful for their commitment and dedication to conducting their research, which has been pivotal in making this book. We cannot express our appreciation enough for their remarkable efforts.

We are grateful for the enthusiasm and support from our colleagues at Pandit Deendayal Energy University, Gandhinagar, Gujarat, India, University of Technology Sydney, Australia, and Victoria University, Melbourne, Australia for their valuable contributions toward the successful completion of this book. Their expertise, guidance, and support were instrumental in shaping the thoughts and ideas, and we are deeply indebted to them for their unwavering commitment and dedication. Their valuable inputs and constructive feedback have helped us to refine the work and make it more comprehensive and informative.

We are grateful to Springer Nature for giving us the opportunity to compile this book and for their expert guidance throughout the process. We would like to thank the production staff who have worked tirelessly to ensure that this book meets the highest standards of quality. We are proud to be associated with Springer Nature and grateful for their continued support in disseminating knowledge and advancing research in our field.

Our gratitude is boundless for the extraordinary reviewers who have devoted their time and attention to closely examining our chapters with their perceptive vision. Their invaluable feedback and comments are deeply appreciated and treasured. We owe an immense debt of gratitude to all those who have supported us and made this dream a reality. Our heartfelt thanks to them as they have put their faith in our endeavor and made it possible for us to bring it to fruition.

Our family members have been there for us every step of the way as we embarked on this exciting project. Each of them has been a source of endless encouragement, patience, and support; for that, we are eternally grateful. Words cannot express the depths of our appreciation for their unwavering commitment. This book was only possible due to the patience that God graciously bestowed upon us. We are grateful to God for His generous blessings.

We have devoted our time, effort, and passion to crafting this book as accurately and error-free as possible. We would be incredibly appreciative if you have any thoughts, comments, or suggestions that could help us elevate this work to the next level.

Dr. Manoj Sahni
Dr. José M. Merigó
Dr. Walayat Hussain

Contents

Editors and Contributors

About the Editors

Dr. Manoj Sahni is working as an Associate Professor and Head at the Department of Mathematics, School of Technology, Pandit Deendayal Energy University, Gandhinagar, Gujarat, India. He has more than eighteen years of teaching and research experience. He holds an M.Sc. degree in Mathematics from Dayalbagh Educational Institute (Deemed University), Agra, M.Phil. from I.I.T. Roorkee, and a Ph.D. degree in Mathematics from Jaypee Institute of Information Technology (Deemed University), Noida, India. He has published more than 80 research papers in peer-reviewed journals, conference proceedings, and chapters with reputed publishers like Springer, Elsevier, etc. He has published five research books with reputed publishers like Springer, Taylor & Francis and others. He also serves as Reviewer for many international journals of repute. He is also an editorial board member of many reputed journals. He has conducted the first, second, and third International Conferences on Mathematical Modeling, Computational Techniques, and Renewable Energy in 2020, 2021, and 2022, respectively. He has contributed to the scientific committee of several conferences and associations. He has delivered many expert talks at the national and international levels. He has organized many seminars, workshops, and short-term training programs at PDEU and various other universities. He has also organized a special Symposium at an International Conference (AMACS2018) on Fuzzy Set Theory: New Developments and Applications to Real-Life Problems held in London, UK, in 2018. He is a Member of many international professional societies, including the American Mathematical Society, SIAM, MAA, IEEE, Forum for Interdisciplinary Mathematics, Indian Mathematical Society, IAENG, and many more.

Dr. José M. Merigó (Highly Cited Researcher in Computer Science) is Full Professor at the School of Systems, Management, and Leadership at the Faculty of Engineering and Information Technology at the University of Technology Sydney and Part-Time Full Professor at the Department of Management Control and Information Systems at the School of Economics and Business at the University of Chile. Previously, he was Senior Research Fellow at the Manchester Business School, University of Manchester (UK), and Assistant Professor at the Department of Business Administration at the University of Barcelona (Spain). He holds a Master's and a Ph.D. degrees in Business Administration from the University of Barcelona. He also holds a B.Sc. and M.Sc. degrees from Lund University (Sweden). He has published more than 400 articles in journals, books, and conference proceedings, including journals such as Information Sciences, European Journal of Operational Research, IEEE Transactions on Fuzzy Systems, Expert Systems with Applications, International Journal of Intelligent Systems, Applied Soft Computing, Computers & Industrial Engineering, and Knowledge-Based Systems. He has also published several books with Springer and with World Scientific. He is on the editorial board of several journals including Applied Soft Computing, Technological and Economic Development of Economy, Journal of Intelligent & Fuzzy Systems, International Journal of Fuzzy Systems, Kybernetes and Economic Computation, and Economic Cybernetics Studies and Research. He has also been Guest Editor for several international journals, Member of the scientific committee of several conferences, and Reviewer in a wide range of international journals. Recently, Thomson & Reuters (Clarivate Analytics) has distinguished him as Highly Cited Researcher in Computer Science (2015–2017). He is currently interested in decision-making, aggregation operators, computational intelligence, and bibliometrics and applications in business and economics.

Dr. Walayat Hussain is the Head of the Discipline of Information Technology at Peter Faber Business School, Australian Catholic University, Australia. He has an academic and industry experience of more than 18 years. Before joining the ACU, he was a Lecturer at Victoria University Melbourne, Lecturer and a Post-doctoral Research Fellow at the University of Technology Sydney for seven years. He worked as an Assistant Professor and postgraduate program coordinator at the department of Computer Science at BUITEMS University for many years. He holds a Ph.D., Master, Postgraduate Diploma, and Bachelor's (Hons.) degrees in Computing and Information Systems. Walayat's research areas are Service Computing, Business Intelligence, AI, Information Systems, Computational Intelligence, Machine Learning, and Decision Support Systems. He has contributed to theory and application and developed a new approach for making an optimal informed decision in complex systems. He has published more than 80 high-quality research document in various top-ranked ERA-A*, A, ABDC- A, JCR/SJR Q1 journals. He is currently an Associate Editors for the IET- Communications and Forecasting, international journals, and has been guest editors for more than 25 international journals. He has served as the General Chair. Co-General Chair in multiple international conferences. He is a fellow of European Alliance for Innovation. He has won multiple national and international research awards and recognitions. He is the recipient of the prestigious National

Research Award by the Government of Oman 2021, the VUBS Award for Excellence in Research 2022, the Best Paper Award at IoTaaS Australia 2021, the Best Paper Awards at 3PGCIC 2015, 2016 Poland, South Korea, and FEIT HDR Publication Award by the UTS Australia, 2016.

Contributors

Kirti Aggarwal Jaypee Institute of Information Technology, Noida, India

Gastelum Chavira Diego Alonso Universidad Autonoma de Occidente, Culiacan SIN, Mexico

Pavel Anselmo Alvarez Department of Management and Economic Sciences, Universidad Autónoma de Occidente, Culiacan, Mexico

Anuja Arora Jaypee Institute of Information Technology, Noida, India

Joel Azzopardi Department of AI, Faculty of ICT, University of Malta, Msida, Malta

Denisse Ballardo-Cárdenas Universidad Autónoma de Occidente, Culiacán, Sin., México

Maria Bernal Universidad Tecnologica de Culiacan, Los Angeles, Culiacan Rosales, Mexico

Fabio Blanco-Mesa Escuela de Administración de Empresas, Facultad de Ciencias Económicas y Administrativas, Universidad Pedagógica y Tecnológica de Colombia; Av. Central del Norte, Tunja, Colombia

Diego Alonso Gastélum Chavira Universidad Autónoma de Occidente, Culiacán, Sin., México

Nishant Doshi Department of Computer Science, Pandit Deendayal Energy University, Gandhinagar, India

Alejandra Duarte Department of Engineering and Technology, Universidad Autónoma de Occidente, Culiacan, Mexico

Dianny Fernandez-Samacaa Escuela de Administración de Empresas, Facultad de Ciencias Económicas y Administrativas, Universidad Pedagógica y Tecnológica de Colombia; Av. Central del Norte, Tunja, Colombia

Tania Gadea School of Medicine, Universidad Autónoma de Sinaloa, Culiacan, Mexico

Pedro Ángel García Aguirre Departamento de ingeniería Industrial y Manufactura, Universidad Autónoma de Ciudad Juárez, Chihuahua, Mexico

Erwin Martínez Gómez Departamento de ingeniería Industrial y Manufactura, Universidad Autónoma de Ciudad Juárez, Chihuahua, Mexico

Pinkimani Goswami University of Science and Technology Meghalaya, Techno City, India

Mahendra Kumar Gourisaria School of Computer Engineering, KIIT Deemed to Be University, Bhubaneswar, Odisha, India

Ana Laura Herrera-Prado Universidad Autónoma de Occidente, Mazatlan, Sinaloa, Mexico

Sandali Khare School of Computer Engineering, KIIT Deemed to Be University, Bhubaneswar, Odisha, India

Gogula Suvarna Kumar Department of Computer Science and Engineering, MVGR College of Engineering, Vizianagaram, Andhra Pradesh, India

Veronica Larreta Department of Management and Economic Sciences, Universidad Autónoma de Occidente, Culiacan, Mexico

Ernesto León-Castro Faculty of Economics and Administrative Sciences, Universidad Católica de La Santísima Concepción, Concepción, Chile;
Business School, Tecnológico de Monterrey, Culiacán Rosales, Sin, México;
Unidad Navojoa, Instituto Tecnologico de Sonora, Navojoa, Sonora, México

David Luviano-Cruz Departamento de ingeniería Industrial y Manufactura, Universidad Autónoma de Ciudad Juárez, Chihuahua, Mexico

Smruti Malhar Mahapatro KIIT School of Management, KIIT University, Bhubaneswar, Odisha, India

R. Maheswari School of Computer Science Engineering, Vellore Institute of Technology, Chennai, India

N. Mohana School of Advanced Sciences, Vellore Institute of Technology, Chennai, India

Aishwarya Mohanty KIIT School of Management, KIIT University, Bhubaneswar, Odisha, India

Manuel Muñoz-Palma Management Department, Universidad de Sonora, Hermosillo, Sonora, Mexico

Lingam Naveen Biju Patnaik Institute of Information Technology and Management Studies (BIITM), Bhubaneswar, India

Elizabeth Olmos-Martínez Universidad Autónoma de Occidente, Mazatlan, Sinaloa, Mexico

Ayush V. Patel School of Computer Engineering, KIIT Deemed to Be University, Bhubaneswar, Odisha, India

Chintan Patel Department of Computer Science, University of Sheffield, Sheffield, UK

Luis Pérez-Domínguez Departamento de ingeniería Industrial y Manufactura, Universidad Autónoma de Ciudad Juárez, Chihuahua, Mexico

Diego Quintero-Avellaneda Escuela de Administración de Empresas, Facultad de Ciencias Económicas y Administrativas, Universidad Pedagógica y Tecnológica de Colombia; Av. Central del Norte, Tunja, Colombia

Akshika Rastogi Department of Statistics, D. N. College, Meerut, India

Andrés Alejandro Fernández Rosas Escuela de Administración de Empresas, Facultad de Ciencias Económicas y Administrativas, Universidad Pedagógica y Tecnológica de Colombia; Av. Central del Norte, Tunja, Colombia

Manoj Sahni Department of Mathematics, School of Technology, Pandit Deendayal Energy University, Gandhinagar, India

Abu Saleh Canberra Business School, University of Canberra, Canberra, ACT, Australia

Sandra Samuel School of Computer Science Engineering, Vellore Institute of Technology, Chennai, India

Kenia Sanchez-Valenzuela Universidad Autónoma de Occidente, Culiacán, Sin., México

Madan Mohan Singh North-Eastern Hill University, Shillong, Meghalaya, India

R. B. Singh Department of Statistics, D. N. College, Meerut, India

S. R. Singh Department of Mathematics, C.C.S. University, Meerut, India

Surbhi Singhal Department of Statistics, Vardhaman College, Bijnor, India

Rabi N. Subudhi KIIT School of Management, KIIT University, Bhubaneswar, Odisha, India

Siba Kumar Udgata School of Computer and Information Sciences, University of Hyderabad, Hyderabad, India

Delia-Julieta Valles-Rosales Department Chair-Industrial Management & Technology, Frank H. Dotterweich-College of Engineering Texas A &M University-Kingsville, Kingsville, Texas, USA

Deyanira Velázquez Department of Management and Economic Sciences, Universidad Autónoma de Occidente, Culiacan, Mexico

Arwa H. Zabian Faculty of Science and Information Technology, Jadara University, Irbid, Jordan

Chapter 1
Internet of Things-Based Digital Life to Provide Assisted Living for Visually Challenged Persons

Siba Kumar Udgata and Gogula Suvarna Kumar

1 Introduction

Humans are not disabled. A person can never be broken. Our built environment, our technologies, is what is broken and disabled. We the people need not accept our limitations, but can transfer disability through technological innovation. According to an estimate of World Health Organization (WHO) Prevention of Blindness, about 285 million people are visually impaired throughout the world; where 39 million people are blind and 246 million have low vision (severe or moderate visual impairment [1, 2]).

The onset of vision loss or blindness changes an individual's life significantly. As the individual deals with the loss—its severity with relationship, identity, and functioning become clear. Visually challenged and impaired people always depend on others for their locomotion. They become dependent on other people for performing their day-to-day activities also. As a result, previously balanced relationships may take on a new aspect of dependency and can make people feel isolated and homebound. Even, within home, colliding with furniture, fear of tripping, falling down, dropping things, etc. can lead those with vision impairment to restrict their movements and activities, becoming more sedentary—and thereby putting both mental and physical health at risk.

Mobility of the visually challenged persons can be described as the flexibility to move with safety and ease in their environment whenever they want, without relying

S. K. Udgata
School of Computer and Information Sciences, University of Hyderabad, Hyderabad 500046, India
e-mail: udgata@uohyd.ac.in

G. S. Kumar (✉)
Department of Computer Science and Engineering, MVGR College of Engineering, Vizianagaram, Andhra Pradesh, India
e-mail: suvarna.cse@mvgrce.edu.in

M. Sahni et al. (eds.), *Novel Developments in Futuristic AI-based Technologies*, Algorithms for Intelligent Systems, https://doi.org/10.1007/978-981-99-3076-0_1

on another person. The most commonly used mobility aids are canes and guided dogs. However, a cane provides a limited preview for the user and as a result, the user has to be more careful while walking and so has to move very slowly with caution. For guiding dogs, the training and coordination of the dog with the visually challenged person is quite challenging and difficult. The results in this case are also minimal.

The increasing number of visually challenged people to a higher level demands the understanding and assistance to their needs. According to a study published by Lancet Global Health (2017), the number of people in the world who are visually challenged is all set to increase by at least three times by the year 2050 in spite of the fact that the percentage of visually challenged people with that of the total population is steadily decreasing owing to the better healthcare facilities. The same research further elucidates the decline in the visually challenged people per population group from 0.75% in 1990 to 0.48% in 2018, and the rate of moderate to severe visual impairment from 3.83% to 2.90% over the same period of time owing to improved health interventions [1].

Further, it is to be noted that the rate of decline of treated vision cases is unmatched by the rapid growth in blindness. Although reasonable work is being put into enhancing the health services, it is fair to say that the untreated people are leftover without significant heed paid to improve their current quality of life. Equipping the blind with the right technology can empower them as well as gain time for medical professionals to reach a solution/cure.

2 Review of Status of Research and Development

Over the years, the development of assistive techniques to guide the visually impaired is evident. This exploration helps empower the low-vision populace toward independence enabling free movement in their environment without guidance from others.

In 2005, Yuan et al. [3] have discussed the virtual white cane sensing device based on active triangulation that can measure distance at a rate of 15 m/s. Besides measuring distances, this device can detect surface discontinuities, such as the foot off a wall, a step, or a drop-off. This is obtained by analyzing the range of data collected as the user swings the device around tracking planar patches and finding discontinuities.

In 2013, a "Smart walking stick" presented by Rana and Sayemil [4] studied obstacle detection by a ping sensor and the obstacle distance is communicated to the visually impaired by the vibration of the motor. The model is based on a ping sensor (for detecting an obstacle), wet electrode, vibration motor, and the buzzer. Further, that same year, in a navigation and monitoring system for the visually challenged, an ARM7 controller and ultrasonic sensing technology were used to detect the obstacle and inform the distance of the obstacle to the person who is visually impaired. GSM and GPS technologies for localization for the person who is visually impaired are also included in the work.

In 2014, a 3D ultrasonic stick was developed to assist visually challenged individuals. The stick uses an ultrasonic sensor to detect obstacles in three directions: front, left, and right. It also includes a vibration motor that varies in intensity depending on the distance of the obstacle. The stick is equipped with advanced technologies, including GSM and GPS modules, voice recognition, and voice synthesis.

The design process consists of two main parts. Firstly, the obstacle detection and voice generation unit is designed using ultrasonic detection and voice synthesis technologies. Secondly, the localization and monitoring unit is designed using GPS and GSM technology. These two units are then combined to create the complete device.

Recently Facebook in joint collaboration with Ray Ban developed wearable Facebook glass which can play sound, capture the image and video, pair with other Bluetooth devices [5]. There are other companies like Bose also developed a wearable eye glass for audio system. Google smart glass also uses some of the above-mentioned technologies to develop devices for better viewing and experience. But, none of these products are actually targeted at visually challenged persons. We also proposed in the past some work for the benefit of the challenged people using sensor network and Internet of Things [6–9].

In this work, we empathize with the pain and suffering of the visually challenged people and try to develop a device and method using the Internet of Thing technology. This device will have a digital inclusive life for visually challenged people to help them in leading the day-to-day life in a better and confident manner.

2.1 Background of the Problem

The interest and desire to build a commodity for the sightless is increasing among the researchers and technology enthusiasts. In most of the methods proposed in the literature for the visually challenged people, accuracy and efficiency is mediocre. Further, these devices and methods are void of a user-friendly and accessible interface.

Despite the global effort to develop a sustainable system, the pragmatism is doubtful. Furthermore, none of the methods delivers a consolidation of various features that can render a persistent guide unnecessary, nor does it present a real-world perception for the visionless.

3 The Objectives

All physically challenged people including visually challenged ones do not like to be assisted by others as it brings down their morale. They wish to lead a life as normal as possible. Any help in this direction makes them feel inclusive and confident.

The fundamental objective of this work is to integrate computer vision, cognitive abilities, real-world perception, and audio representation into a unique wearable prototype device with the purpose of assisting visually challenged people to help them navigate indoors as well as outdoors without any active assistance from another person. The main objective is to assist the visually challenged people in such a manner that they feel empowered and lead a normal life in a comfortable and confident manner.

To achieve the fundamental objectives, we set the research objectives as follows:

(a) To design a wearable prototype that is comfortable and easy to access by the visually challenged
(b) To consolidate features that can enhance the virtue of life of the challenged
(c) To develop a cheap, affordable, and efficient way to help the visually challenged to navigate with greater comforts, speed, and confidence
(d) To ensure the visually challenged safety while moving around and accomplishing daily activities.

3.1 The Research Objective and Digital Life Prospective

In the ancient times "Buddha" preaching is based on four important objectives, viz., (i) Duhkha, (ii) Samudaya, (iii) Nirodha, and (iv) Marg. The research objectives with a human touch also focus on (i) Empathize, (ii) Analyze, (iii) Solution, and (iv) Test and Use. The relationship with the preaching and research objectives is shown in Table 1. In the Buddha preaching, 'Duhkha' means suffering, "Samudaya" means "Source", "Nirodha" means protect, and "Marg" means path.

In research objectives, the following analogy can be brought in.

(i) in the empathize stage the researcher tries to identify the pain points and feels it
(ii) in the Analyze phase the research finds the source of such problem or pain points
(iii) in the Solution phase, we try to find a solution for protecting against the source and pain points
(iv) and finally "Test and use" phase denotes the testing of the proposed solution and then putting it to use.

Table 1 Buddha preaching and research objective analogy

Buddha preaching	Research objective
Duhkha	Empathize
Samudaya	Analyze
Nirodha	Solution
Marg	Test and use

3.2 Proposed System Design

Technology has always helped human surpass their limitations and move toward a better life. Even for the visually challenged, the need for an effective solution that can help them perform daily activities/chores is definitely desirable. In this present work, extensive has been put in to overcome the difficulties in the existing methods and to provide a cost-effective and user-friendly system for the visually challenged.

The proposed solution involves designing a wearable as an aid for the visually challenged to cope with the day-to-day challenges they face. Efforts to use the proposed system are minimal—neither does the system require special training nor does it include holding a huge device for a long distance. The intent of this research, further, solicits the need for a light, portable and, most importantly, affordable device. On a larger scale, with improvements in the prototype, the cost will reduce and drastically benefit the society.

The electronic guidance system, which is proposed as a constructive assistant and support, can function accurately, hardly requires any training, is easy to wear and operate. The design is a special wearable device that can be worn like spectacles and is based on a Raspberry Pi controller. This device is equipped with hardware equipment like Pi camera, GPS module, and Bluetooth headphones.

The proposed system mainly has six assistive features to enhance the digital life of the challenged person in their day-to-day life.

1. Object Detection: The challenged person will be updated about the surrounding objects. This enables the user to understand his/her environment to take corresponding and corrective actions.
2. Know Your Friend: This proposed system is designed to be smart enough to detect basic attributes and recognize friends. This allows the challenged person to greet their friends in a usual manner.
3. Text Reader: This feature is essentially an image-to-voice converter that keeps the user informed about elemental activities like reading price tags, reading labels on medicine.
4. Geo-fencing: This functionality monitors the geographical position of the challenged person, and they will be alerted with buzzers/vibration in case they go out beyond the desired distance or in case of a detour. Also, the guardians are alerted and can live-track the user.

Fig. 1 Block diagram of the proposed system

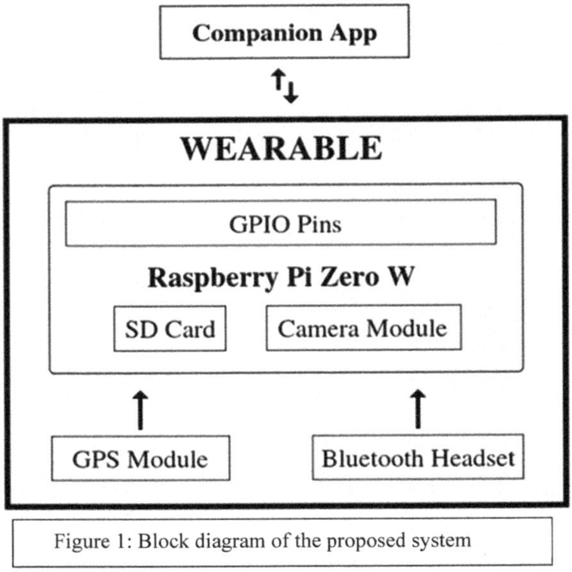

5. Community App for helping the challenged person: The proposed system also contains a mobile App which is designed to build a community around its users. It helps them to connect among themselves and also with other persons in the society.

The first three features belong to the broad domain of computer vision, the fourth feature belongs to the domain of location-based services using GPS system, and the last feature belongs to the domain of community development App. We will discuss the detailed features, their technical details for implementation, and use cases in the following section Design Methodologies, Implementation, Results, and Discussions.

The block diagram of the proposed system is given in Fig. 1.

4 Design Methodology, Implementation, Results, and Discussions

An electronic guidance system is proposed with the following design objectives.

- as a constructive assistant and support
- can function accurately
- hardly requires any training
- is easy to wear and operate (worn like spectacles).

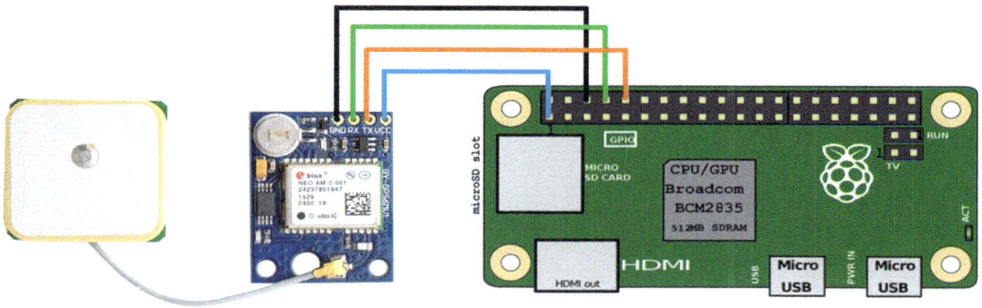

Fig. 2 The integrated architecture of the wearable device with Raspberry Pi Zero W and other modules

An integrated architecture of the proposed wearable spectacle device with Raspberry Pi Zero W and other modules is shown in Fig. 2. It also shows the abstract view of the connections between the modules.

4.1 Computer Vision-Based Features

This feature of the electronic guidance systems consists of three features (1) Object Detection, (2) Know your friend, and (3) Text reader that will assist the user to lead a normal life and make them feel confident in their day-to-day lives. All the three features use the same hardware (with three different software modules to be triggered by the trigger button provided). These services use the Bluetooth module embedded in the device and the paired earphone to communicate the results in the audio format to the user.

The main hardware components utilized for these above features are:

- Raspberry Pi zero controller
- Pi Camera
- Buttons (three buttons for triggering).

The three buttons in the designed wearable device act as the trigger for a corresponding feature. In order to initiate any feature, we capture the image, which has to be converted into a voice directive using the Pi camera. The image will then be stored in the memory module of the Raspberry Pi. In the proposed model, the image will be uploaded onto the cloud and the resultant URL will be passed for corresponding analysis using the REST API of Microsoft computer vision application. After the analysis, the objects in the image will be added as tags.

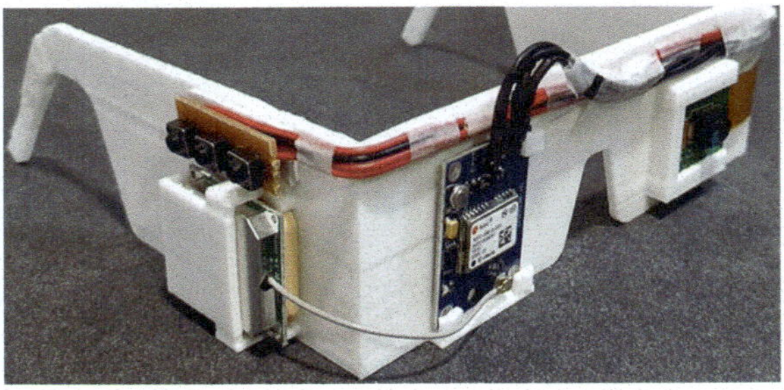

Fig. 3 The designed wearable device with GPS module

4.1.1 Object Detection

The 'object detection' is a lucid feature that updates the challenged person about the surrounding objects based on the research results reported in the literature [10–12]. This enables the user to understand his/her environment to take corresponding actions. The spectacle-shaped device developed for the purpose (as shown in Fig. 3) captures an image of the surrounding using an imager camera. By analyzing the visual content of the image, the system can ascertain the objects in the captured image. Further, the images can be tagged into categories to help the user comprehend the surrounding environment and the user can have a feel of the situation around. Based on those tags, we can generate a 'description" in human understandable format. The description will then be passed to a voice assistance that will inform the user of the surrounding objects.

Technology used

Microsoft's Computer Vision API helps us detect the visual content of an image. Using this, we can ascertain the objects in the captured image.

Further, the images can be tagged into categories to help the user comprehend the atmosphere and situation around him/her. The code snippet for the object detection feature is shown in Table 2.

A sample image of a room having a laptop computer and chairs in a room where the experiment is conducted is shown in Fig. 4. Using the proper trigger button, the camera image is captured by the spectacle-like wearable device. This image is then analyzed by the software code (the code snippet is shown in Table 2).

The execution window of the Python code developed for the purpose is shown in Fig. 4. Please note the output description (encircled with red color) which says that "a desk with a computer and chair in the room" is detected in the room.

4.1.2 Know Your Friend

The electronic guidance system developed is made to be smart enough to detect basic attributes and determine your friends. Based on the earlier research reported in the literature [13–16], we developed this system as an integrated and user-friendly model

Table 2 The code snippet for 'object detection' feature

```
def obj_det():
   speak("Object Detection Activated")
   cam = picamera.PiCamera()
   cam.capture('image.jpg')
   cam.close()
   img = upload_files()
   headers = { 'Ocp-Apim-Subscription-Key' :
'*****' }
   api_url = 'https://westcentralus.api.cogni-
tive.microsoft.com/vision/v2.0/describe?maxCandi-
dates=1'
   response= requests.post(api_url,headers=head-
ers,json={'url': img})
   res=response.json()
   speak(res['description']['cap-
tions'][0]['text'])
```

Fig. 4 A sample image of the surrounding in a room for object detection feature

for the visually challenged persons. This allows the challenged person to greet their friends in the usual manner without feeling any discomfort. To detect the friends of the user, we initially create an image database to store the information about a set of friends of the user along with their names. When the right trigger button is used to capture the image of the friend on his/her arrival, the image of the friend is captured and uploaded to the cloud for analysis. The analysis for this feature includes individual comparison of the current image with the set of images in the database based on the derived image features. Here also we use the different features of the REST API to detect and extract the features of the image captured (Fig. 5).

The proposed model can alert the user about the following:

- Facial expression of the person

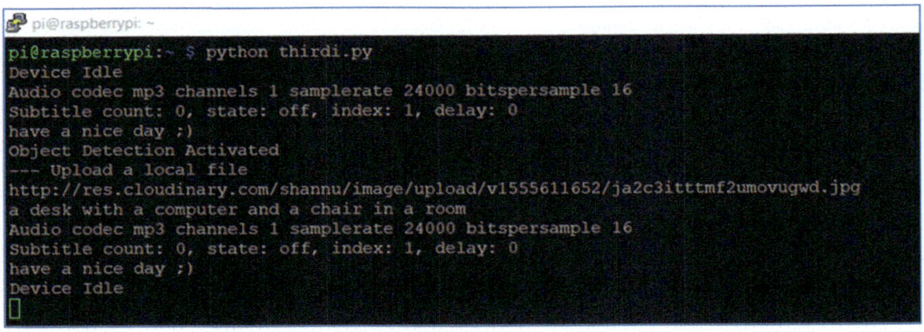

Fig. 5 The execution window of the Python code of 'Object detection' feature

- Age and sex of the person
- Tell you the name (if an acquaintance)
- Other physical details like the color of the shirt.

Using a captured image, we can determine the age and gender of a particular person. The face is stored in the database, and it is able to identify the person in front of the user. If not, it provides a range of facial and physical attributes. Also, it recognizes the person's expressions, thus, stating his/her mood to the challenged person. This can help build a genuine interaction between the two.

Technology Used

Using Microsoft's Face API, we can determine the age and sex of a person. If the face is stored in the database, it identifies the person in front of our user. If not, it provides a range of facial and physical attributes. The code snippet for this activity "know your friend" is given in Table 3.

Table 3 The code snippet for 'Know your friend' feature

```
def verify_faces(id1,id2):
   headers = { 'Ocp-Apim-Subscription-Key' :
'****' }
   face_api_url = 'https://westcentra-
lus.api.cognitive.microsoft.com/face/v1.0/ver-
ify'
   response= requests.post(face_api_url,head-
ers=head-
ers,json={'faceId1':id1,'faceId2':id2})
   res=response.json()
   if(res['isIdentical']==True):
       return(True)
   else:
       return(False)
```

Fig. 6 A sample image of the surrounding in a room for 'Know your friend' feature

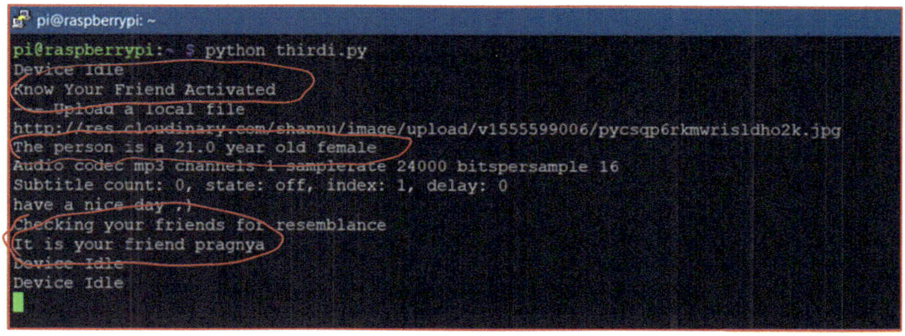

Fig. 7 The execution window of the Python code of feature 'Know your friend'

A sample image of a friend approaching the challenged person in the room is shown in Fig. 6. Using the proper trigger button (button-2), the camera image is captured by the spectacle-like wearable device. This image is then analyzed by the software code (the code snippet is shown in Table 3).

The execution window of the Python code developed for the purpose is shown in Fig. 7. Please note the output description (encircled with red color) which says that "The person is a 21 years old female", "It is your friend Pragnya" are detected in the room.

4.1.3 Image to Text Reader

This feature is, essentially, an image-to-voice converter that keeps the user informed about elemental activities like

- Reading price tags
- Reading labels on medicine.

To convert the image into text, we have used Microsoft Cognitive Service Computer Vision API which extracts text using Optical Character Recognition (OCR) from an image into a human readable character stream [17, 18]. To analyze the image, we can either upload an image or specify the URL of the image. If required, the OCR corrects the inclination of the image (rotation and size invariant functions are used) text to recognize it and provides the frame coordinates of each word. It enables the recognition of printed and handwritten text in an image. Computer Vision can detect and extract both printed and handwritten text from images of various objects with diverse surfaces and backgrounds. Currently, printed and handwritten text recognition is available only for English.

Technology Used

Using optical character recognition (OCR), we can detect the text in images. Image Processing with Computer Vision API (Microsoft) helps us analyze visual data and extract embedded text.

The custom-made voice assistant helps us convey the text to the user. The code snippet for this activity "Image to text reader" is given in Table 4.

A sample image of an image with text to be read by the challenged person is shown in Fig. 8. Using the proper trigger button (button-3), the camera image is captured by the spectacle-like wearable device. This image is then analyzed by the software code (the code snippet is shown in Table 4).

Table 4 The code snippet for 'Image to text reader' feature

```
headers = {'Ocp-Apim-Subscription-Key': "****"}
   params  = {'mode': 'Handwritten'}
   data    = {'url': img}
   response = requests.post("https://westcentra-
lus.api.cognitive.microsoft.com/vision/v2.0/recog-
nizeText", headers=headers, params=params,
json=data)
```

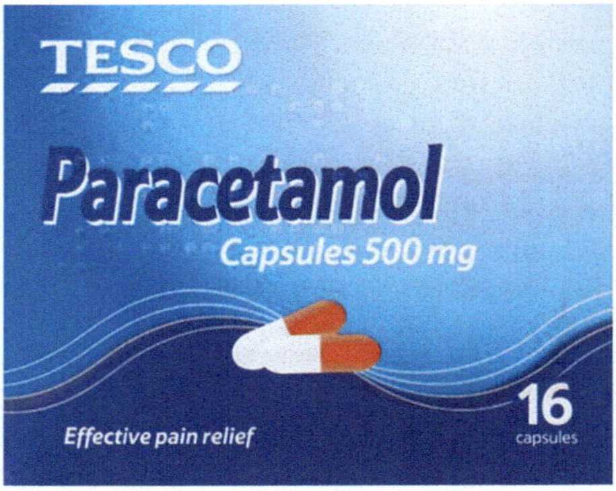

Fig. 8 A sample image of 'Image to text reader' feature

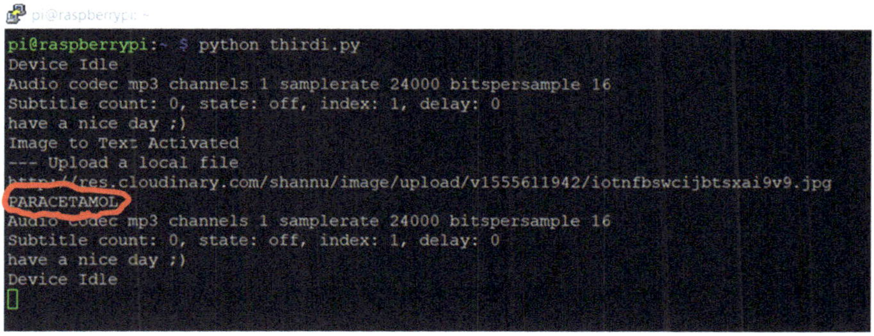

Fig. 9 The execution window of the python code of feature "Image to text reader" feature

The execution window of the Python code developed for the purpose is shown in Fig. 9. Please note the output description (encircled with red color) which says that a "Paracetamol" is detected in the image.

4.2 Geo-Fencing

Please note that the first paragraph of a section or subsection is not indented. The first paragraphs that follow a table, figure, equation, etc. do not have an indent, either.

Subsequent paragraphs, however, are indented.

This is a location-based service having a feature that envisions the creation of a virtual geographic boundary using GPS module [19–21]. It monitors their (challenged persons) position and the challenged person will be alerted with buzzer/vibration in case they go out beyond the desired distance or in case of a detour. Also, the guardians/caregivers are alerted and can live-track the user. Establishment of the 'geofence' is done via a web App by providing the location coordinates (latitude and longitude) and the necessary radius of the enclosure. Once the geo-fence is set up, we can track the person every second using the GPS module present in the wearable spectacle-like device of the user. The GPS module constantly sends the location coordinates to the server, where, we verify if the location is within the radius of the geo-fence or not? If the location coordinates are in bounds, there is no problem. However, if the given location is out of bounds, then a notification will be sent to the user and their caretaker.

Technology Used

Using the Google Maps platform and U-blox GPS module, we can record and live-track to guardians the location of the challenged person. We built a webpage using HTML to set a Geofence. Using REST API, we keep track of the location of the user.

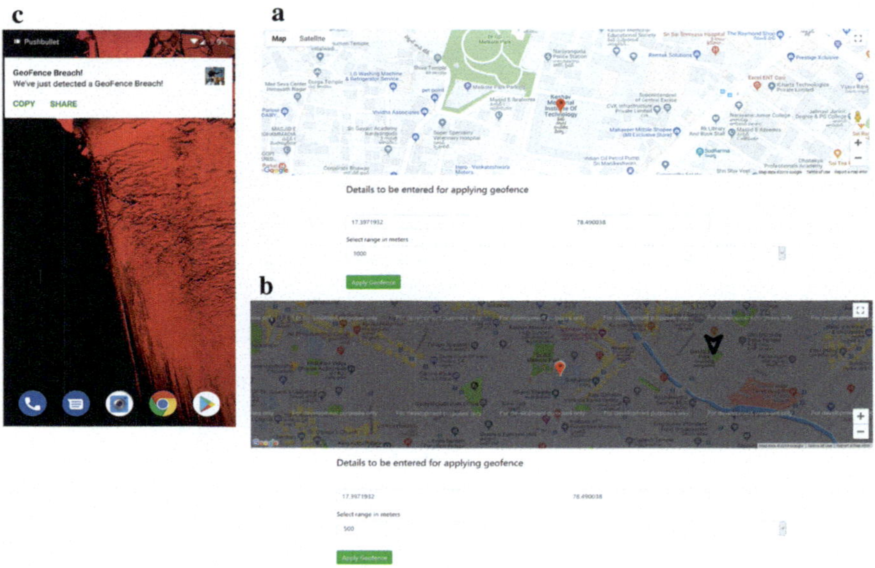

Fig. 10 **a** A sample image of 'Geofence' setting feature. **b** A sample image of 'Geofence' setting feature. **c** A sample image of 'Geofence' breach feature

In case of geo-fence breach, both, the user and their guardians are alerted. The notification is sent using PushBullet API.

Figure 10 shows the details of the 'Geofence' implementation. Figure 10a, b show the interactive screen to enter the geofencing boundaries. This geo-fence boundary is set by the caregiver for the challenged person to constantly monitor the challenged person. Figure 10c shows an example of the screen where it shows when a geo-fence is breached. This message is being forwarded to the caregiver also along with a vibration and buzzer.

4.3 Community Building

The prosed electronic guidance system intends to build a community around its users [22, 23]. It helps them connect among themselves and with others. It allows volunteers to be online/alert to help the visually challenged people. This way, anyone can help the challenged in mere seconds anytime, anywhere. In return of their services, volunteers are rewarded with Social Credit Points (SCP). The users can identify people (similar to them) around them who are seeking assistance and spend time with like-minded people. The user can also discover friends through the developed mobile application.

Technology Used

Fig. 11 The interaction windows of the mobile App for 'Community development' feature

Using the Flutter framework for an application, we provide an interface for commoners to volunteer and help the ones in need. The users can send requests when they require assistance. The users can connect with others like them and socialize. We use MLab services for request transactions and notifications. The screens and interactions are designed keeping the visually challenged person in view which requires minimal interface and also does not need any visuals. The interface normally has tap (double tap) on the screen and voice interactions.

Figure 11 shows the interaction windows for the challenged person to create a request for some services in the interaction window with limited or no visual interface. Figure 12 shows the interaction window of the caregivers/friends who are willing to accept the request and provide the requested services. The total social credit points accumulated by the requester are also displayed in the middle window.

5 Conclusions and Future Scope

In today's world, disability of any kind can be hard and it is the same in case of blindness. Blind and visually challenged people are generally left underprivileged and offering vision to them is difficult many times in medical terms. Any services to improve their quality of life in an inclusive manner will be a step forward to improve their life. Technological inclusion in their digital life in a careful and systematic manner will help them in a great way.

In this paper, we have presented a IoT-based wearable device for assisted living in order to enhance the digital life of visually challenged persons. The work is aimed

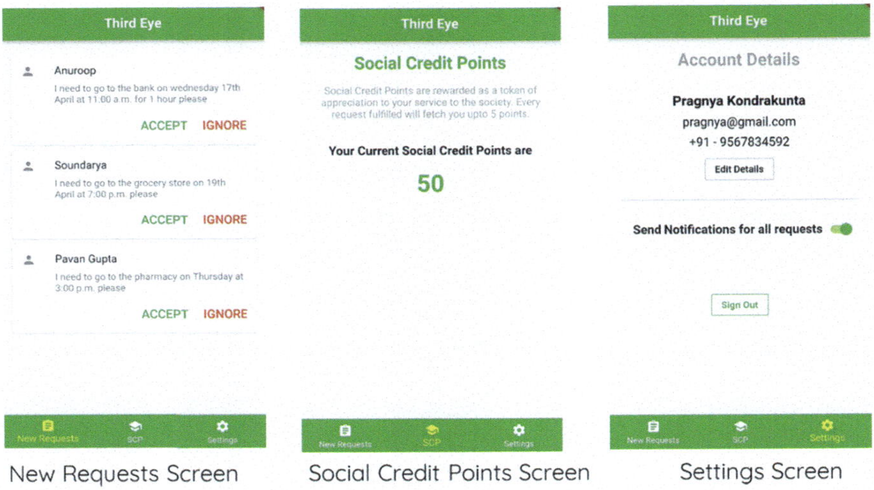

Fig. 12 The interaction windows of the mobile App for 'Community development' feature

at designing a system for physically challenged persons to improve their digital life and assist them in their day to life to increase their confidence and quality of life.

We proposed mainly four applications embedded in a single device to assist the visually challenged people to know about the environment in which he/she is working, to recognize a friend with different attributes so that they will comfortable and confident in greeting them, to read a text in the image to help them to find out the price tag, expiry date and other details without the help of anybody else. These functionalities help them to live their life as a normal person (with the assistance of the wearable device).

In future we shall mainly focus on making the developed device more optimized and user-friendly for the users to use it. We are also thinking of integrating the developed device with smartphones to have low latency. We can also try to integrate this service with the existing third-party applications like Uber/Ola/Swiggy/Zomato, etc. to extend their services to the physically challenged persons in a seamless manner. The same application can be integrated with robotics to improve the services of the robots.

Acknowledgements The author would like to acknowledge the help of Mrutyunjay Anuroop and Pragnya Kondrakunta, the student interns who helped to develop the system and test its efficacy.

References

1. https://data.worldbank.org/indicator/SP.POP.65UP.TO?locations=IN
2. https://www.populationpyramid.net/india/2050/
3. Yuan D, Manduchi R (2005) Dynamic environment exploration using a virtual white cane. In: 2005 IEEE computer society conference on computer vision and pattern recognition (CVPR'05), ,vol 1, pp 243–249. https://doi.org/10.1109/CVPR.2005.136
4. Mahmud MH, Saha R, Islam S (2013) Smart walking stick—an electronic approach to assist visually disabled persons. Int J Sci Eng Res 4(10)
5. https://www.ray-ban.com/usa/discover-ray-ban-stories/clp
6. Udgata SK, Sindhura G, Sabat SL (2009) A novel architecture for intelligent and adaptive wireless sensor network based alarming system. In: 2009 fifth international conference on wireless communication and sensor networks (WCSN), pp 1–6. https://doi.org/10.1109/WCSN.2009.5434805
7. Udgata SK, Kumar KP, Sabat SL (2010) Swarm intelligence based Resource Allocation Algorithm for cognitive radio network. In: 2010 first international conference on parallel, distributed and grid computing (PDGC 2010), pp 324–329. https://doi.org/10.1109/PDGC.2010.5679975
8. Udgata SK, Suryadevara NK (2021) COVID-19, Sensors, and Internet of Medical Things (IoMT). In: Internet of things and sensor network for COVID-19. Springer Briefs in Applied Sciences and Technology. Springer, Singapore. https://doi.org/10.1007/978-981-15-7654-6_3
9. Udgata SK, Suryadevara NK (2021) Advances in sensor technology and IoT framework to mitigate COVID-19 challenges. In: Internet of things and sensor network for COVID-19. Springer Briefs in Applied Sciences and Technology. Springer, Singapore. https://doi.org/10.1007/978-981-15-7654-6_4
10. Ess A, Leibe B, Schindler K, Van Gool L (2008) A mobile vision system for robust multi-person tracking. In: 2008 IEEE conference on computer vision and pattern recognition, pp 1–8. https://doi.org/10.1109/CVPR.2008.4587581
11. Pathak AR, Pandey M, Rautaray S (2018) Application of deep learning for object detection. Procedia Comput Sci 132:1706–1717, ISSN 1877-0509,. https://doi.org/10.1016/j.procs.2018.05.144
12. Kane SK, Morris MR (2017) Let's Talk About X: combining image recognition and eye gaze to support conversation for people with ALS. In: DIS '17: Proceedings of the 2017 conference on designing interactive systems, pp 129–134. https://doi.org/10.1145/3064663.3064762
13. Dehghan A, Ortiz EG, Shu G, Masood SZ (2007) DAGER: deep age, gender and emotion recognition using convolutional neural network. CoRR abs/1702.04280
14. Chen D, Cao X, Wen F, Sun J (2013) Blessing of dimensionality: high-dimensional feature and its efficient compression for face verification. IEEE Conf Comput Vis Pattern Recognit 2013:3025–3032. https://doi.org/10.1109/CVPR.2013.389
15. Prieto LAB, Oplatkova ZK (2017) A performance comparison of two emotion-recognition implementations using OpenCV and Cognitive Services API. In: MATEC web of conferences, vol 125, p 02067. https://doi.org/10.1051/matecconf/201712502067
16. Turabzadeh S, Meng H, Swash R, Pleva M, Juhar J (2018) Facial expression emotion detection for real-time embedded systems. Technologies 6:17
17. Hubbell D (2016) Seeing AI: new technology research to support the blind and visually impaired community. In: Microsoft build conference
18. Zhao T, Ryan P (2018) Making sense of hand written text in scanned documents using Azure cognitive services. In: Microsoft build conference
19. Rahate SW, Shaikh MZ (2016) Geo-fencing Infrastructure: location based service. Int Res J Eng Technol (IRJET) 03(11)
20. Rai P, Lakhmani R, Singh Y, Varshney P (2018) Security based alert system using geofencing. IJARIIE 4(2)
21. Basid SA, Nuraida PM, Tolle H, Ramdani F (2017) Designing module E-complaint system based on geotagging and geofencing. Int J Interact Mob Technol (IJIM) 11(3) (2017)

22. Brinkley J, Hoffman D, Tabrizi N (2017) A social networking site portable profile system for blind and visually impaired users based on cloud and semantic web technologies. IEEE Int Conf Cogn Comput (ICCC) 2017:104–111. https://doi.org/10.1109/IEEE.ICCC.2017.21
23. Reis A, Paulino D, Filipe V, Barroso J (2018) using online artificial vision services to assist the blind—an assessment of microsoft cognitive services and google cloud vision. In: Rocha Á, Adeli H, Reis L, Costanzo S (eds) Trends and advances in information systems and technologies. WorldCIST' 18 2018. Advances in intelligent systems and computing, vol 746. Springer, Cham. https://doi.org/10.1007/978-3-319-77712-2_17

Chapter 2
Impact of Information and Communication Technology on Tourist Spending

Ana Laura Herrera-Prado, Elizabeth Olmos-Martínez, and Ernesto León-Castro

1 Introduction

Tourism is one of the socio-economic activities that, over time, has presented a constant and growing evolution at the international level, except for the last three years, due to the unusual health events that have occurred [41]. Likewise, the incursion of Information and Communication Technologies (ICT) in bringing tourists closer to tourist destinations is becoming more important and indispensable every day [3]. This context has directly influenced the era of globalization in tourism, which is characterized by the increase in the competitiveness of tourism companies, causing a positive impact of technologies in all areas of the social and economic life of the countries [15]. These new conditions that arise promote the development of both Smart Cities (CI and Smart Tourist Destinations (DTI) [8, 26]).

In this sense, using ICTs worldwide has grown by leaps and bounds. People in the twenty-first century want to stay informed and communicated 24 h a day, 365 days a year [6, 27]. In addition, this situation has increased due to the confinement measures taken in the face of the current pandemic caused by Covid-19. As a result, many people use their cell phones, laptops, tablets or computers to carry out many daily activities [2]. However, even though the pandemic has negatively impacted the development of tourist activities since people's mobility has had to be reduced, it is also a motivation for ICTs to be used as a tool that helps to take steps that can avoid contact between people and thus reduce the chances of contagion [11]. This

A. L. Herrera-Prado · E. Olmos-Martínez (✉)
Universidad Autónoma de Occidente, 82100 Mazatlan, Sinaloa, Mexico
e-mail: eolmosm.udo@gmail.com

E. León-Castro
Business School, Tecnológico de Monterrey, 80100 Culiacan Rosales, Sinaloa, Mexico

M. Sahni et al. (eds.), *Novel Developments in Futuristic AI-based Technologies*, Algorithms for Intelligent Systems, https://doi.org/10.1007/978-981-99-3076-0_2

situation encourages the tourism sector to increasingly include new information and communication technologies in its management strategies.

Over the years, both tourism and the use of ICT in the different economic sectors have shown an upward trend [21]. However, it is essential to verify quantitatively if there is a relationship between these two events to be sure that if tourism companies invest in technological resources, the tourism sector will increase tourist spending and thereby increase the competitiveness of tourist destinations. Considering the above, the objective of this article is to determine the ratio of tourist spending at the international level based on ICT indicators included in the travel and tourism competitiveness index. For this, it is important to mention that the World Economic Forum has a travel and tourism competitiveness index (TTCI, for its acronym in English Travel & Tourism Competitiveness Index) that allows for evaluating the competitiveness of tourist destinations [42].

Finally, this work was structured into four sections. The first deals with the theoretical framework, where the importance of tourism worldwide is presented, as well as the evolution of ICTs, to understand the incorporation of ICTs in the tourism industry. The second section explains the methodology used in this research; the third section shows the results and discussion. Finally, conclusions and suggestions are presented.

2 Theoretical Framework

2.1 Importance of Tourism Worldwide

According to the World Tourism Organization [41], the panorama of international tourism showed positive figures until before the Covid-19 pandemic. The number of international tourist arrivals in the world that was registered in the year 2019 was 1,461 million tourists. This represents a variation of +3.8% compared to 2018. Where of the five UNWTO regions, the Middle East registered the highest percentage of variation, observing +7.6%, followed by the Asia and the Pacific region with an increase of 4.6% and by Africa with a variation of +4.2%. Then comes Europe with a variation of +3.7% and the Americas with 2%. Travel and tourism are key sectors for economic development and job creation worldwide. The World Travel and Tourism Council [43] reported that this sector represented 10.4% of world GDP and generated 319 million jobs, 10% of total employment in 2018. In addition, travel and tourism was the second fastest growing sector in the world in this same year, with +3.9%, ahead of health services (+3.1%), information and communication technologies (+1.7%) and services financial (+1.7%). Its growth was only inferior to the manufacturing sector, which grew by 4%.

France, Spain and the United States are the countries in the top three positions worldwide according to the number of tourist arrivals they received in 2019 (90.2, 83.7 and 79.3 million, respectively). On the other hand, considering income from tourism, the same three countries are in the top positions but in a different order, with the United States in the first place, Spain in second and France in third (214.1, 83.8 and 68 billion dollars respectively) [13].

Tourism is an economic activity that in many countries contributes significantly to the economy of regions, towns, or states, generating well-being in different sectors of society. Several authors have published information highlighting tourism's positive impact in the economies where it is practiced. This shows a direct and positive relationship between tourism dynamics and the economic growth of tourist destinations [18, 35, 36].

2.2 ICT Evolution

The information and communication technologies 'consist of information technologies, as well as telecommunications, broadcast media, all kinds of audio and video processing and transmission, and network-based control and monitoring functions' [40], p. 15. Likewise, the Federal Telecommunications Institute defines ICT as 'the technological convergence of computing, microelectronics and telecommunications to produce information in large volumes, to consult it and transmit it over enormous distances. It encompasses all those technologies that make up the Information Society, such as, among others, computing, the Internet, multimedia or telecommunications systems' [23], p. 5.

Technology has been used for communication since the dawn of civilization, for example, with the use of smoke signals, the mail, Morse code and early telephone calls; the beginning of the evolution of ICTs will be considered the construction of the first digital computer, called ENIAC, and publicly announced on 14 February 1946 [30]. The next step in the evolution of ICT is the creation of computer networks, such as the current Internet. The first of these wide coverage networks was the ARPANET, created in 1969 to communicate research centres, being the direct antecedent of the World Wide Web. Since the creation of the ENIAC, computers have been used to make weather models, air traffic simulators, conference systems, mathematical systems, experimental databases and even a computerized chess player. However, each computer was isolated, and their communication was very complicated. It was not until the creation of the ARPANET that computers from different manufacturers could establish a long-distance connection on a single network [1].

In addition to the first networks between computers, the first cell phone prototype was also worked on in the 1970s. In 1969, the Bell System invented the first commercial cellular radiotelephone, demonstrating multiple train passengers making telephone calls at speeds of 100 miles per hour on a 450 MHz frequency. In the same year, Motorola implemented the first transistor cell phone, quickly replacing radio cell phones on the market, although still of great weight and size. However, in November 1971, Intel created the first microprocessor, the 4004, which allowed Motorola in 1973 to patent its own radio cell phone system and its first handheld cell phone prototype [19]. After the ARPANET came a new global computer network, the

World Wide Web, created by Tim Berners-Lee in 1989, a British scientist working for CERN (for its French acronym Conseil Européen pour la Recherche Nucléaire). Built to share scientific information between universities, just like its predecessor, the ARPANET, the World Wide Web became the most widely used network because, in April 1993, CERN made the World Wide Web public domain software. And later, it would release as a free license to encourage its spread [10].

Likewise, the union of cell phones and the Internet allowed the creation of smartphones, thanks to microprocessor technologies. In August 1994, the IBM Simon went on sale, the first smartphone or smartphone, although at that time it was not called that, it contained many of the functions of current smartphones, such as a calendar, notepad, sending messages and e-mails and phone calls [4]. Currently, the development of microprocessor technologies allows the design of smartphones that are more powerful, faster, lighter and cheaper, becoming one of the most widely used communication devices with the computer and the landline telephone. Furthermore, due to the improved performance of these ICTs, both smartphones and computers and the Internet, social networks were created, digital media in which people interact immediately with people thousands of kilometres away. From Facebook, YouTube, TikTok, WhatsApp, Instagram and Twitter, the various social networks instantly unite people from all over the world with common interests. Companies, particularly tourist companies such as hotels and airlines, have used these social networks since the creation of the Internet to increase their market and improve their profits [28]. Tourism companies can use social networks in various ways in public sites and private areas, from advertising and marketing, hotel room purchase and rental service or purchase of airline tickets and market research [20].

According to [32], ICTs and the Internet are the pillars of the technological innovations that society is experiencing, causing a revolution in their use and originating a new culture of communication and marketing. This has generated that more people are in contact with the Internet and their social networks daily, taking advantage of all the benefits they offer. Promoting companies, including those in the tourism sector, to implement the use of these technological tools so that they can continue to be competitive [9].

2.3 Incorporation of ICT in the Tourism Industry

Tourist destinations face the challenge imposed by this new trend where developing cutting-edge technological infrastructure requires a significant weight [25]. Because this technical solution must guarantee the eco-efficiency of the destination, improve the connection between destinations, people and services and create emotional connections between the visitor and the destination, among other aspects [7]. Some examples of ICT applied to tourism are free Wi-fi in tourist areas, applications for Smartphones, mobile payments, QR codes, Cloud computing, augmented reality, Internet of Things (IoT), Big Data, video guides, devices intelligent optics that facilitate historical immersion processes, personalized experiences in museums, etc. [7].

For this reason, tourism companies are adopting the use of ICTs to improve their operations' administration and satisfy and facilitate the services that tourists require. In hotels and restaurants, for example, implementing these tools is observed to increase the competitiveness of tourist destinations [16]. Likewise, the wi-fi connection is one of the most requested services by tourists since it is convenient for tourist sites to have an Internet connection facility so that tourists can carry out their online marketing transactions. In other words, the importance of online payments has increased among travellers in the last decade due to the facilities presented with the use of technologies in the tourism sector. Many of the transactions that tourists carry out when they want to make a trip are carried out through online operations [29].

The evolution of tourism, the use of ICTs and the irruption of the Internet have favoured the emergence of a new tourist profile widely adapted to technologies [17]. This explains the need for tourist destinations to adopt technological solutions such as the development of mobile applications [27]. This type of software is mainly aimed at providing tourist information, promoting and geolocating the tourist offer of the destination and making it easier for tourists to share their travel experiences. However, as can be seen, the usefulness of these applications is very varied, so tourist destinations must channel them towards tourist services that are least offered and most demanded by travellers [24].

Likewise, the use of mobile devices to manage trips is increasing daily. Smartphones are displacing computers due to these devices' greater comfort and ubiquity [38]. As this technology's use for transactions by tourists increases, mobile payments also increase, achieving greater satisfaction in their tourist experience. Since travellers can make payments with their cell phones from anywhere, they are with an internet connection, these payments can be directed towards lodging, transport or any tourist service [33].

Another technological tool used in tourist destinations is the Quick Response (QR) barcode, a two-dimensional barcode that links to a unique address that identifies a person. Web page on the Internet (URL, for its acronym in English for Uniform Resource Locator) requires mobile devices and QR code reader applications. It promotes and commercializes tourist products and services [37], allowing tourists to have an agile interaction with the destination by facilitating off/online communication and achieving greater satisfaction in their travel experience, generating possible loyalty on the part of the traveller. Regarding the Internet of Things (IoT), it is applied more to sensors connected to the Internet that allows specific areas of the destination to be monitored in real-time—for example, managing to identify the more saturated sites, allowing the tourist to be able to choose other options to improve their tourist experience [12].

3 Methodology

To verify the impact of ICT on tourist spending, quantitative research was carried out using the simple linear regression method and the multiple linear regression method as statistical tools. To do this, five indicators were selected that were related to information technologies and included in the World Economic Forum's travel and tourism competitiveness index. In addition, 50 countries were selected from a total of 266, considering that they had complete information for each nation for the years 2017 and 2018. To do this, the data for each indicator was searched in the database—data from the Travel and Tourism Competitiveness Report (TTCR) [42].

The statistical models used in this research were structured by the independent variable called 'ICT' and by the dependent variable 'Tourism'. On the one hand, for the independent variable, the following indicators were considered: (1) Digital demand for cultural and entertainment tourism; (2) Use of ICT for transactions between companies (electronic marketing); (3) Use of ICT for business-to-consumer transactions (customer relationship management); (4) Individuals who use the Internet and (5) Mobile network coverage rate. On the other hand, tourist spending was used as an indicator for the dependent variable.

In the first stage, individual analysis of the Simple Linear Regression models of tourist spending was carried out based on each ICT indicator. Outliers or atypical values were previously detected using the Hi (leverage) techniques [39], Cook's Distance [14] and DFITS Statistics [5] to eliminate said atypical data. Likewise, data normality tests were performed to transform those data that did not meet this parametric criterion. Then, in a second stage, a Multiple Linear Regression analysis was performed between tourist spending and all ICT indicators combined. Finally, a Stepwise Regression or step-by-step function was conducted to find a new regression model to explain better the relationship between tourist spending and the significant ICT indicators in this analysis.

4 Results

4.1 First Stage: Results of the Simple Linear Regression Analysis

Outlier detection was done through Hi (leverage), Cook's Distance and DFITS. To detect outliers or atypical values that could produce deviated results in this research, three tests were applied, in which the Hi, Cook's Distance and DFTIS Statistics techniques were used. To do this, the expenditure of tourists (Y) registered in 50 countries in 2017 and 2018 was considered, in addition to the data corresponding to the same countries and years, concerning the following ICT indicators: Digital demand for cultural and entertainment tourism (X1), ICT use for business-to-business transactions (X2), Internet use for business-to-consumer transactions (X3), Individuals using the Internet (X4) and Mobile network coverage rate (X5).

In the first test, the Hi (leverage) technique was carried out; in the results, no negative data were obtained. However, in the second test, called the Cook Distance technique, four pieces of data were observed that presented an atypical behaviour (data 10, 48, 60 and 98), which correspond to the spending of tourists from China and the US in the 2017 and 2018 (Fig. 1).

In the third test, using the DFITS Statistical technique, four negative data were also obtained, which coincided with the four atypical data of the previous test. These data belong to the United States and China in 2017 and 2018. Therefore, when obtaining the same atypical data in two of the three tests, we eliminated them from the database used in this work (Fig. 1: data 10, 48, 60 and 98).

Evaluation of Simple Linear Regression Models for each ICT indicator regarding tourist spending. It is important to mention that a Kolmogorov–Smirnov normality test was performed on each of the study variables, a statistical significance test used to determine if the population from which a sample was drawn has a normal distribution, according to what was suggested by [34]. The results of this test showed that the only variables with a normal distribution are X2 and X3. Therefore, the variables Y, X1, X4 and X5 had to be transformed using the natural logarithm function to use the parametric analysis of Linear Regression. As a result of this transformation, the variables Y, X1, X4 and X5 were renamed Yt, X1t, X4t and X5t to carry out the corresponding normality tests. The graphs of the residuals shown in Fig. 2 show that after having made the transformations to the variables that required it, the distributions for each of the relationships represented correspond to a normal distribution.

In the five normal probability plots presented in Fig. 2, it is observed that most of the residuals are close to the normal probability line. Regarding the graphs of

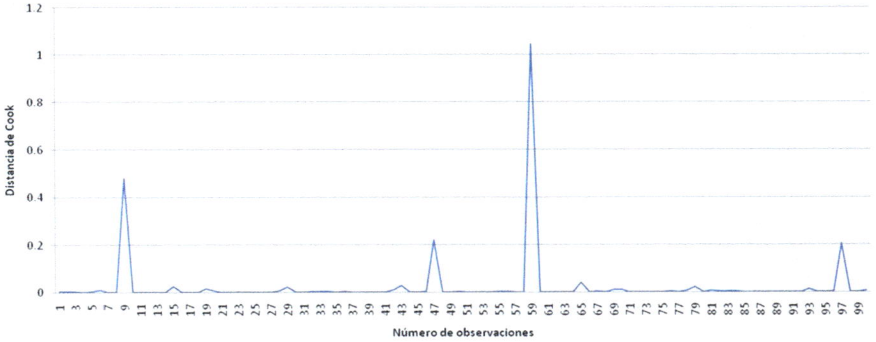

Fig. 1 The plot of Cook's Distance for the complete model* *Of tourist spending (Y) based on all the ICT indicators studied: Digital demand for cultural and entertainment tourism (X1), Use of ICT for business-to-business transactions (X2), Internet use for business-to-business transactions consumer (X3), Individuals who use the Internet (X4) and Mobile network coverage rate (X5).

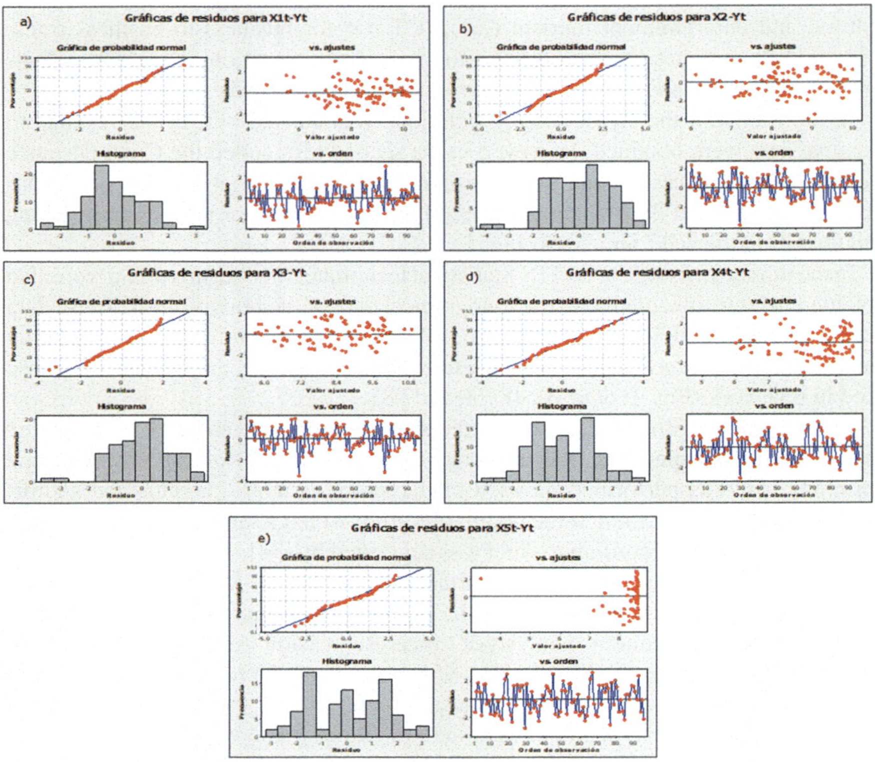

Fig. 2 Residual graphs for the relationships between tourist spending (Yt) and the ICT indicators studied* ***a*** Digital demand for cultural and entertainment tourism (X1t), **b** Use of ICT for business-to-business transactions (X2), **c** Internet use for business-to-consumer transactions (X3), **d** Individuals who use the Internet (X4t) and **e** Mobile network coverage rate (X5t)

Residuals versus Fits, we see that the residuals are spread out at the top and bottom of the zero line. Likewise, in the graphs of Residuals vs Order, they present the same behaviour. Finally, in most histograms, a homogeneous distribution is observed, tending to normal. This allows validation of the application of simple linear regression analysis.

Figure 3 shows the fitted line graphs to show the behaviour of the dependent variable with each of the five independent variables. The points that represent the relationship between tourist spending (on the ordinate axis) and the ICT indicators individually (abscissa axis) are observed in each graph. The continuous line represents the equation of the Simple Linear Regression model for each pair of variables indicated in the graph, whose equation is also presented for each case. In four of these graphs, a good fit of the data to the model obtained is observed, except for the relationship between tourist spending and the mobile network coverage rate, where the observed data are biased to the right of the graph.

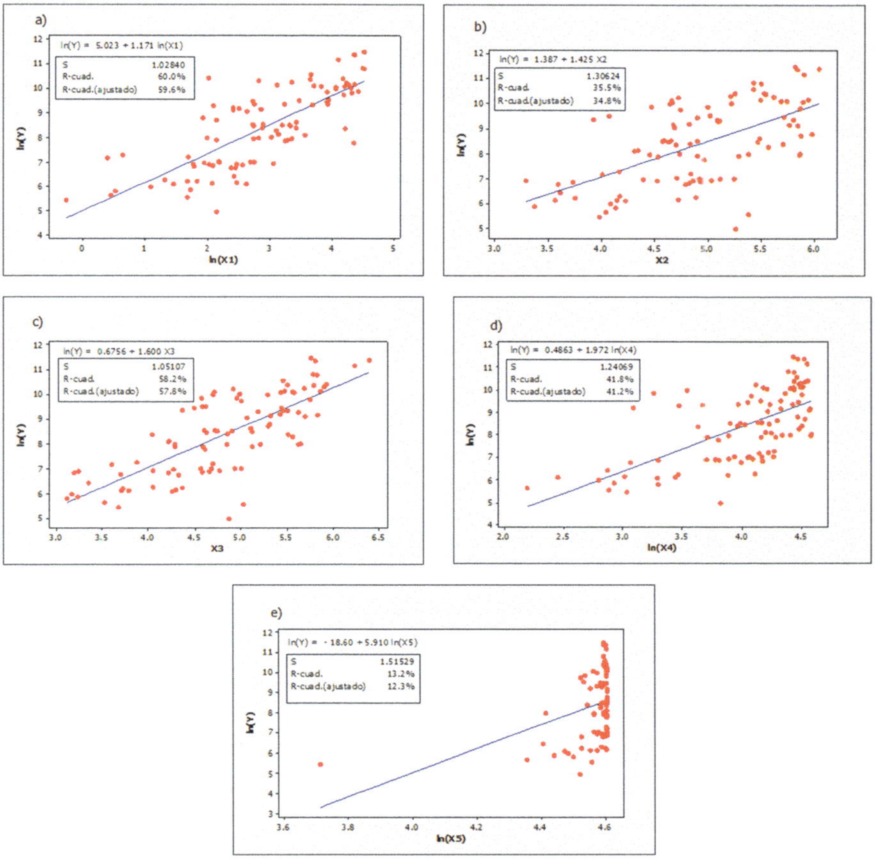

Fig. 3 Fitted Line Graphs for the relationships between tourist spending (Y) and the ICT indicators studied* ***a*** Digital demand for cultural and entertainment tourism (X1), **b** Use of ICT for business-to-business transactions (X2), **c** Internet use for business-to-consumer transactions (X3), **d** Individuals who use the Internet (X4) and **e** Mobile network coverage rate (X5)

Table 1 shows the results of the Simple Linear Regression analysis and the equations relating to tourist spending based on each ICT indicator studied. In all the models, a positive relationship is shown because tourist spending is observed when increasing the values of each ICT indicator (Fig. 3), which is evidenced by the positive sign of the coefficients of X in the corresponding equations. In addition, each ICT indicator was a significant variable for their respective model since the P value in each case is less than 0.05.

The coefficient of determination (R2) of the models presented in Table I is the proportion of the total variance of tourist spending (Y) explained by each of the corresponding regression equations. These results show that according to the ICT

Table 1 Results of the Simple Linear Regression Analysis*

ICT indicator	Model	Coef	Coef of EE	T	P	VIF	R^2 (%)	R^2 adjusted (%)
X1	ln(Y) = 5.02 + 1.17 ln(X1)	Constant 5.02 ln(X1) 1.17	0.3031 0.0985	16.57 11.88	0.000 0.000	1.000	60.0	59.6
X2	ln(Y) = 1.39 + 1.43 X2	Constant 1.39 X2 1.43	0.9840 0.1981	1.41 7.19	0.162 0.000	1.000	35.5	34.8
X3	ln(Y) = 0.68 + 1.60 X3	Constant 0.68 X3 1.60	0.6832 0.1398	0.99 11.45	0.325 0.000	1.000	58.2	57.8
X4	ln(Y) = 0.49 + 1.97 ln(X4)	Constant 0.49 ln(X4) 1.97	0.9712 0.2399	0.50 8.22	0.618 0.000	1.000	41.8	41.2
X5	ln(Y) = −18.60 + 5.91 ln(X5)	Constant −18.60 ln(X5) 5.91	7.1390 1.5620	-2.61 3.78	0.011 0.000	1.000	13.2	12.3

*For the models of tourist spending (Y) and each of the ICT indicators studied: Digital demand for cultural and entertainment tourism (X1), Use of ICT for transactions between companies (X2), Use of the Internet for transactions from business to consumer (X3), Individuals using the Internet (X4) and Mobile network coverage rate (X5)

indicator related to tourist spending, the adjustment of the corresponding model can be between 13.2% and 60% of its variance. Each model can be used to calculate tourist spending from the related ICT indicators, with an associated uncertainty ranging from 86.8% to 40%, depending on the case.

4.2 *Second Stage: Results of the Multiple Linear Regression*

As a result of the Multiple Linear Regression analysis, the regression equation presented below was obtained. In it, tourist spending was used as a dependent variable, in this case, represented as Y. As independent variables, the five ICT indicators were included (digital demand for cultural and entertainment tourism (X1), use of ICT for business-to-business transactions (X2), use of the Internet for business-to-consumer transactions (X3), individuals who use the Internet (X4) and mobile network coverage rate (X5)). As a result, this complete model's coefficient of determination (R2) is 79.3%, which is higher than all obtained through the Simple Linear Regression analysis. Therefore, this model can calculate tourist spending with an uncertainty of 20.7% from all ICT indicators studied.

$$\ln(Y) = 4.98 + 0.690\ln(X1) - 1.05X2 + 1.59X3 + 0.667\ln(X4) - 0.819\ln(X5) \quad (1)$$

Table 2 shows the results of the Multiple Linear Regression analysis for the complete model of tourist spending based on all the ICT indicators studied. In this model, positive relationships are observed between tourist spending and digital demand for cultural and entertainment tourism (X1), Internet use for business-to-consumer transactions (X3) and individuals using the Internet (X4). While the use of ICT for transactions between companies (X2) and the mobile network coverage rate (X5), the relationships are negative. Most of these ICT indicators turned out to be significant variables for this model, except for the mobile network coverage rate (X5) since the P value of this ICT indicator was greater than 0.05.

The variance analysis results for this complete Multiple Linear Regression model reflect the statistical significance of the relationship between tourist spending and the five ICT indicators since the P value of this equation is well below 0.05 (Table 3). Additionally, the Durbin-Watson statistical test was conducted to verify no autocorrelation between the study variables [22]. An alpha of 0.05 was used, with n equal to 96 observations and k equal to 6 variables, obtaining a lower limit of 1.535 and an upper limit of 1.802. Therefore, the value of the Durbin-Watson statistic for this regression equation is 1.89412, and this value is greater than the upper limit because there are no related errors. Therefore, it is concluded that there is no autocorrelation between the variables.

Subsequently, a Stepwise Regression or step-by-step function was performed to find a new regression model that better explains the relationship between tourist spending and the five ICT indicators studied. For this, 96 observations were included, and alpha to enter and an alpha to withdraw of 0.15 were used. The results of this process are shown in Table IV, where four steps were carried out, through which it was obtained that if only four of the five ICT indicators are included, a regression equation is obtained with an R2 value of 79.16% with a P value less than 0.05 in the four independent variables. Therefore, it can be stated that the digital demand

Table 2 Results of the multiple linear regression analysis*

ICT indicator	Coef (β)	Coef of EE	T	P	VIF
Constant	4.98	3.9230	1.27	0.207	
X1	0.69	0.0945	7.30	0.000	1.705
X2	−1.05	0.2721	−3.87	0.000	5.639
X3	1.59	0.2616	6.08	0.000	6.778
X4	0.67	0.2148	3.11	0.003	2.161
X5	−0.82	0.9103	−0.90	0.371	1.366

*For the complete model of tourist spending (Y) based on all the ICT indicators studied: Digital demand for cultural and entertainment tourism (X1), Use of ICT for transactions between companies (X2), Use of the Internet for business-to-consumer transactions (X3), Individuals using the Internet (X4) and Mobile network coverage rate (X5)

Table 3 Analysis of variance for multiple linear regression*

Variation Source	Degrees of freedom	SC	MC	F	P
Regression	5	197.323	39.465	69.14	0.000
Residual error	90	51.371	0.571		
Total	95	248.694			

*From the complete model of tourist spending (Y) based on all the ICT indicators studied

for cultural and entertainment tourism (X1), the use of ICTs for business-to-business transactions (X2), the use of the Internet for business-to-consumer transactions (X3) and the individuals who use the Internet (X4) are significant variables for this new model (Table 4).

After identifying the significant ICT indicators through Stepwise Regression, a Multiple Linear Regression analysis was performed for the partial model of tourist spending based on said indicators, obtaining the equation presented below.

Table 4 Stepped regression results (step by step)*

	Step			
	1	2	3	4
Constant	5.023	1.487	2.481	1.502
X1	1.171	0.751	0.71	0.669
T value	11.88	7.62	7.54	7.31
P value	0	0	0	0
X3		0.98	1.77	1.6
T value		7.19	6.64	6.14
P value		0	0	0
X2			−0.95	−1.06
T value			−3.37	−3.89
P value			0.001	0
X4				0.61
T value				2.98
P value				0.004
S	1.03	0.829	0.786	0.755
R-cuad	60.03	74.3	77.13	79.16
R-cuad. (adjusted)	59.6	73.75	76.38	78.24
Cp de Mallows	82.2	22	11.7	4.8

*Of tourist spending (Y) based on the ICT indicators studied: Digital demand for cultural and entertainment tourism (X1), Use of ICT for business-to-business transactions (X2), Internet use for business-to-consumer transactions (X3), Individuals who use the Internet (X4) and Mobile network coverage rate (X5)

$$\ln(Y) = 1.50 + 0.669\ln(X1) - 1.06X2 + 1060X3 + 0.608\ln(X4) \quad (2)$$

In this equation, the dependent variable is tourist spending (Y), while the independent variables included were the digital demand for cultural and entertainment tourism (X1), the use of ICTs for business-to-business transactions (X2), use of the Internet for business-to-consumer transactions (X3) and individuals using the Internet (X4). As a result, this partial model's coefficient of determination (R2) is 79.2%, practically the same as that obtained using the complete model. Therefore, this partial model can also be used to calculate tourist spending, with an uncertainty of 20.8%, from the four ICT indicators that were significant in the Stepwise Regression analysis.

Table 5 shows the results of the Multiple Linear Regression analysis for the partial model of tourist spending based on four significant ICT indicators resulting from the Stepwise Regression. In this partial model, positive relationships are observed between tourist spending and digital demand for cultural and entertainment tourism (X1), Internet use for business-to-consumer transactions (X3) and individuals using the Internet (X4). And only a negative relationship was obtained with the use of ICT for transactions between companies (X2). For this partial model, the four ICT indicators included turned out to be significant variables since their P value is less than 0.05.

The results of the analysis of the variance of the partial Multiple Linear Regression model reflect the statistical significance ($P < 0.05$) of the relationship between tourist spending and the four ICT indicators identified by Stepwise Regression (Table 6). The value of the Durbin-Watson statistic for this model is 1.89627. Using an alpha of 0.05, with n equal to 96 observations and k equal to 5 variables, it yielded a lower bound of 1.557 and an upper bound of 1.778; whereas the calculated value is greater than the upper limit, it is considered that there are no errors that are related, therefore, it is concluded that there is no autocorrelation between the variables.

Table 7 presents the Regression Models obtained to estimate tourist spending based on the ICT indicators studied. The first two equations were obtained by Simple

Table 5 Results of the multiple linear regression analysis for the partial model*

ICT indicator	Coef (β)	Coef of EE	T	P	VIF
Constant	1.5025	0.6619	2.27	0.026	
X1	0.6691	0.0915	7.31	0.000	1.602
X2	−1.0580	0.2718	−3.89	0.000	5.638
X3	1.6024	0.2610	6.14	0.000	6.763
X4	0.6076	0.2040	2.98	0.004	1.954

*Of tourist spending (Y) based on the four significant ICT indicators of the Stepwise Regression (step by step): Digital demand for cultural and entertainment tourism (X1), Use of ICT for transactions between companies (X2), Internet use for business-to-consumer transactions (X3), Individuals who use the Internet (X4)

Table 6 Analysis of variance for the multiple linear regression of the partial model*

Variation Source	Degrees of freedom	SC	MC	F	P
Regression	4	196.861	49.215	86.40	0.000
Residual error	91	51.833	0.570		
Total	95	248.694			

*Of tourist spending (Y) based on the four significant ICT indicators resulting from the Stepwise Regression (step by step)

Table 7 Regression models obtained*

Regression equation	R^2 (%)	Analysis
ln(Y) = 5.02 + 1.17 ln(X1)	60.0	SLR
ln(Y) = 0.68 + 1.60 X3	58.2	SLR
ln(Y) = 4.98 + 0.690 ln(X1)–1.05 X2 + 1.59 X3 + 0.667 ln(X4)–0.819 ln(X5)	79.3	MLR
ln(Y) = 1.50 + 0.669 ln(X1)–1.06 X2 + 1.60 X3 + 0.608 ln(X4)	79.2	MLR

SLR = Simple Linear Regression
MLR = Multiple Linear Regression
*To estimate tourist spending (Y) based on the ICT indicators studied: Digital demand for cultural and entertainment tourism (X1), Use of ICT for business-to-business transactions (X2), Internet use for business transactions to consumers (X3), Individuals who use the Internet (X4) and Mobile network coverage rate (X5)

Linear Regression analysis. The first represents tourist spending based on the digital demand for cultural and entertainment tourism, with a coefficient of determination (R2) of 60%. The second equation represents tourist spending based on Internet use for business-to-consumer transactions, with a coefficient of determination of 58.2%. The following equation was obtained through Multiple Linear Regression analysis and represents tourist spending based on the five ICT indicators studied with a determination coefficient of 79.3%. The fourth equation was obtained through Stepwise Regression, which determined that the best model could be represented with four of the five ICT indicators studied. Finally, using the significant ICT indicators from this step-by-step function, the partial model was obtained with a coefficient of determination of 79.2%.

5 Discussion

As [3] comments, the inclusion of ICTs in the tourist's approach to tourist destinations becomes more important and indispensable every day. Proof of this is that the results of the Simple Linear Regression show that the indicator with the greatest influence on tourist spending at the international level is the digital demand for cultural and entertainment tourism, followed by using the Internet for business-to-consumer transactions. This indicates that companies have had to adapt to the globalization of tourism that is characterized by increased competitiveness and a positive social and economic impact, as mentioned [15]. In addition, according to [8, 26] it contributes to the promotion and development of Smart Cities and Destinations. In addition, with less impact but significance, the indicator rates of mobile network coverage coincide with [2]. They comment that many people use cell phones, laptops, tablets or computers to carry out many daily activities.

Regarding the results of the Multiple Linear Regression, it was observed that four of the five indicators that have a joint impact on the spending of tourists at an international level are the digital demand for cultural and entertainment tourism, the use of ICTs for transactions between companies, Internet use for business-to-consumer transactions and individuals who use the Internet. In this regard, the two indicators related to companies show that using technological tools allows them to continue and/or increase their competitiveness [9, 15] market and improve their profits [28] and increase and improve the offer of various services aimed at tourists.

In this sense, all the indicators that impact this multiple model cause a revolution in their use and contribute to a culture of communication and marketing, as pointed out by [32]. First, however, they must guarantee the connection between destinations, people and services to create emotions between the visitor and the destination [7], highlighting the importance of the two significant indicators for tourist spending at an international level. That is directly related to the tourist, such as the digital demand for cultural tourism and entertainment and the individuals who use the Internet; on the latter, the service that tourists request the most is the internet connection [16], so tour operators must be the mediators for the implementation of the said tool in such a way that the demand is covered.

The previous implies that the new tourist profile is widely adapted to technologies [17]. The need for technological solutions in tourist destinations is observed [27], providing tourists with comfort, flexibility, speed, efficiency and connectivity to any type of tourist service [33].

6 Conclusions

Tourist destinations are immersed in the demand for ICT by tourists and visitors, from the most basic (Wi-fi, mobile applications, mobile payments, QR codes, etc.) to the most complex or complementary (Big Data, IoT, augmented reality, intelligent optical devices, etc.) to satisfy and facilitate the requirements and needs of tourists who are widely adapted to technology. All these ICTs contribute to the competitiveness of the tourist destination and are part of the new era of technological solutions. In this sense, it should be remembered that the objective of this article was to determine the ratio of tourist spending at the international level based on ICT indicators included in the travel and tourism competitiveness index. The statistical methods of Simple Linear Regression and Multiple Linear Regression were used to fulfil it. Where data from the years 2017 and 2018 corresponding to 50 countries were considered.

In the first stage of analysis, where the Simple Linear Regression method was used, it was observed that tourists' spending at the international level could be explained by each of the five ICT indicators studied. This means that, in a different proportion, each of these indicators has an impact on tourist spending. For example, the ICT indicator with the greatest influence on tourist spending is the digital demand for cultural and entertainment tourism, followed using the Internet for business-to-consumer transactions. At the same time, the ICT indicator that has the least impact on tourist spending at an international level is the mobile network coverage rate.

On the other hand, according to the results of the Multiple Linear Regression and Stepwise Regression analysis, it was observed that knowing the impact of the ICT indicators studied on the spending of tourists at an international level is better to use only four of the five indicators. TIC. That is, the digital demand for cultural and entertainment tourism, the use of ICT for business-to-business transactions, the use of the Internet for business-to-consumer transactions and the individuals who use the Internet together have a greater impact on spending by consumers. International tourists.

After analysing each of the four models selected in this research, it can be concluded that the best model obtained is the one that represents the spending of tourists (Y), based on the digital demand for cultural and entertainment tourism (X1), the use of ICT for business-to-business transactions (X2), the use of the Internet for business-to-consumer transactions (X3) and the individuals who use the Internet (X4). Because in this partial model, it was verified that all its predictors were significant.

Future research lines can focus on researching the ICT-Tourism relationship in which more indicators are included, both related to ICT and tourism. Research can also be done in specific tourist destinations using these same technological and tourist indicators. Other research could be carried out on smart cities and smart tourist destinations to identify their management and innovation strategies and check their impact on strengthening their competitiveness.

References

1. Abbate JE (1994) From ARPANET to internet: a history of ARPA-sponsored computer networks, 1966–1988. University of Pennsylvania, p 24
2. Anh Dung T, Khanh Linh V, Thi Nog Quy H, Cong Thanh P, Hoang Tien N (2022) ICT application in tourism industry in post-COVID-19 economy in Vietnam. Int J Multidiscip Res Growth Eval 3(1):502–511
3. Bastante Granell V (2018) El Turista 3.0 o Adprosumer: un nuevo reto para el Derecho y la Economía. Universidad de Almería, pp 47–73
4. BBC (2014) World' s first' smartphone' celebrates 20 years. BBC News Mundo, pp 1–3
5. Belsley DA, Kuh E, Welsch RE (2004) Regression diagnostics: identifying influential data and sources of collinearity. Wiley-Interscience, Hoboken, NJ
6. Benckendorff PJ, Xiang Z, Sheldon PJ (2019) Tourism Information Technology, 3rd Edition (3a ed.). CABI

7. Caro Herrero J, Luque Gil A, Zayas Fernández B (2015) Nuevas tecnologías para la interpretación y promoción de los recursos turísticos culturales. Pasos: Revista de Turismo y Patrimonio Cultural 13(4):931–945
8. Castro Alvarez U, González Rodríguez JA, Maldonado Duarte LM (2017) Destinos turísticos inteligentes: ¿estrategia para el desarrollo local en países pobres? Turismo y Desarrollo Local 22. http://www.eumed.net/rev/turydes/22/destinos-turisticos-inteligentes.html
9. Celdrán-Bernabeu MA, Mazón López JN, Giner Sánchez D (2018) Open Data y turismo. Implicaciones para la gestión turística en ciudades y destinos turísticos inteligentes. Revista Investigaciones Turísticas 15:49–78. https://doi.org/10.14198/inturi2018.15.03
10. CERN (2014) The birth of the web|CERN. Cern 1. http://home.web.cern.ch/topics/birth-web
11. Chung N, Lee H, Ham J, Koo C (2021) Smart tourism cities' competitiveness index: a conceptual model. Inf Commun Technol Tour 1:433–438. https://doi.org/10.1007/978-3-030-65785-7
12. Cimbaljević M, Stankov U, Pavluković V (2018) Going beyond the traditional destination competitiveness–reflections on a smart destination in the current research. Curr Issue Tour 22(20):2472–2477. https://doi.org/10.1080/13683500.2018.1529149
13. CNET & CICOTUR (2020) Panorama de la Actividad Turística en México. Año 10(30):45
14. Cook RD (1977) Detection of influential observation in linear regression. Technometrics 19(1):15–18
15. Cornejo JL, Malcolm C (2020) Touristic stakeholders' perceptions about the smart tourism destination concept in Puerto Vallarta, Jalisco, Mexico.Sustainability 12(5):18. https://doi.org/10.3390/su12051741
16. Cruz Estrada I, Miranda Zavala AM (2019) La adopción de las TIC en restaurantes de Puerto Nuevo, Rosarito, Baja California. Innovar 29(72):59–75. https://doi.org/10.15446/innovar.v29n72.77932
17. Duffus Miranda D, Barrera Legorburo L (2018) Modelo Conceptual del Comportamiento de Compra del Turista Digital. EKOTEMAS. Revista Cubana de Ciencias Económicas 4(2):76–91
18. Eyuboglu S, Eyuboglu K (2019) Tourism development and economic growth: an asymmetric panel causality test. Curr Issues Tour 9. https://doi.org/10.1080/13683500.2019.1588863
19. Farley T (2006) Mobile telephone history. Futur Mob Phones 12
20. Fernández-Poyatos MD, Papí-Gálvez N (2017) eTurismo: estudio de criterios de segmentación clásicos del usuario online que compra por internet. ICONO 14, 15(2):168–189. https://doi.org/10.7195/ri14.v15i2.1066
21. Gholipour HF, Andargoli AE, Arjomandi A, Foroughi B (2021) Capital investment in telecommunications infrastructure and tourist arrivals in developing countries: does the public–private sectors relationship matter? Tour Econ. https://doi.org/10.1177/13548166211014814
22. Gujarati D, Porter D (2010) Econometría. Mc Wraw-Hill, México
23. IFT (2019) Uso de las TIC y Actividades por Internet en México. Instituto Federal de Telecomunicaciones, 95. http://calculadoraprob.ift.org.mx/
24. INVAT.TUR (2015) Destinos turísticos inteligentes. Manual operativo para la configuración de destinos turísticos inteligentes. Agencia Valenciana del Turismo. http://invattur.gva.es/estudio/manual-operativo-para-la-configuracion-de-destinos-turisticos-inteligentes/
25. Ivars-Baidal JA, Celdrán-Bernabeu MA, Femenia-Serra F, Perles-Ribes JF, Giner-Sánchez D (2021) Measuring the progress of smart destinations: The use of indicators as a management toolJ Destin Mark Manag 19. https://doi.org/10.1016/j.jdmm.2020.100531
26. Ivars-Baidal JA, Celdrán-Bernabeu MA, Mazón JN, Perles-Ivars ÁF (2019) Smart destinations and the evolution of ICTs: a new scenario for destination management? Curr Issues Tour 1581–1600. https://doi.org/10.1080/13683500.2017.1388771
27. Jovicic DZ (2019) From the traditional understanding of tourism destination to the smart tourism destination. Curr Issue Tour 22(3):276–282. https://doi.org/10.1080/13683500.2017.1313203
28. Leonardi PM, Huysman M, Steinfield C (2013) Enterprise social media: definition, history, and prospects for the study of social technologies in organizations. J Comput-Mediat Commun 19(1):1–19. https://doi.org/10.1111/jcc4.12029

29. Lorente Bayona L, Ruiz Rua A (2019) The importance of online payment on travel and tourism incomes. A cross-country panel data study. Academicus Int Sci J 20:199–222. https://doi.org/10.7336/academicus.2019.20.15
30. Martin CD (1995) ENIAC: press conference that shook the world. IEEE Technol Soc Mag 14(4):3–10. https://doi.org/10.1109/44.476631
31. OMT (2020) Barómetro OMT del Turismo Mundial, Anexo estadístico, Enero 2020. Barómetro OMT Del Turismo Mundial 18(1):1–48. https://doi.org/10.18111/wtobarometeresp.2020.18.1.1
32. Pardo Lizárraga C, Herrera Prado AL, Rochín Armenta JF (2015) Turismo electrónico para docentes universitarios. In: Magaña Carrillo I, Covarrubias Ramírez R (eds) Competitividad, Sustentabilidad, Innovación: logros y retos del turismo. Puertabierta Editores. Universidad de Colima, pp 311–334
33. Quiñonez-Bedón MF, Tapia-Pazmiño JG, Andrade-Naranjo DS (2019) La utilidad de las TIC en el turismo comunitario. Polo Del Conocimiento 4(11):67–92. https://doi.org/10.23857/pc.v4i11.1175
34. Romero Saldaña M (2016) Pruebas de bondad de ajuste a una distribución normal. Enfermería Del Trabajo 6(3):105–111
35. Sánchez López F (2020) Turismo receptivo y crecimiento económico en México: evidencia de largo plazo. Contaduría y Administración 65(2):19
36. Sandoval Cabrera P (2019) Competitividad y Calidad de vida en Destinos Turísticos de México. InterSedes XX(41):129–149. https://doi.org/10.15517/isucr.v20i41
37. Sandoval Pillajo LL, Enríquez Chugá JF, Sandoval Pillajo AL (2021) Las TIC en la educación y difusión de un producto turístico en Ibarra. Revista Conrado 17(78):291–296
38. Saura J, Palos-Sanchez P, Reyes-Menendez A (2017) Marketing a traves de aplicaciones moviles de turismo. Int J World Tour 4(8):1–12. http://institucionales.us.es/ijwtourism/ojs/index.php/IJWTOURISM/article/view/56
39. Stevens JP (1984) Outliers and influential data points in regression analysis. Psychol Bull 95(2):334–344
40. Tamilselvan N, Sivakumar N, Sevukan R (2012) Information and communications technologies (ICT). Int J Libr Inf Sci 1(1):15–28. https://doi.org/10.4324/9780203717424
41. UNWTO (2021) World tourism barometer. World Tour Barom 19(1):42. https://doi.org/10.18111/wtobarometereng.2021.19.1.110.18111/wtobarometereng.2021.19.1.1
42. World Economic Forum (2019) The Travel & Tourism Competitiveness Report 2019. In Tourism. World Economic Forum
43. WTTC (2019) Travel & Tourism: global economic impact & trends 2019. World Travel & Tourism Council, 20. https://www.statista.com/statistics/233223/travel-and-tourism--total-economic-contribution-worldwide/
44. LNCS Homepage, http://www.springer.com/lncs. Last accessed 21 Nov 2016

Chapter 3
A Sustainable Approach to Manage the Inventory During Pandemic Through Passive RFID Tags

S. R. Singh, R. B. Singh, Akshika Rastogi, and Surbhi Singhal

1 Introduction

For the survival of any organization or business, it is very important to maintain an inventory. During the time of a pandemic, it is the responsibility of the producer to manage the inventory very well so that all the customers will get the item whenever and wherever it is needed.

Radio frequency identification (RFID) is not a new technology in general, but it can be said that it is now in the commercial field. Many organizations are using RFID to maintain their inventory level. RFID tags are not even very expensive for producers to maintain. This technology is very dependable and so much error-free. RFID was discovered during Second World War. Countries like America, Britain, Japan, and Germany were using these technologies to track airplanes approaching them which proved to be a supportive technology at that time. In many other fields such as logistics, medicals, academics, and medical, RFID was continuously used which attracted the attention of researchers, academicians, and scientists. Many organizations have used RFID technology in their own way like many of the companies used RFID tags in their system as anti-theft which is continuously used till now. RFID has proved to be remarkable in supply chain management also.

There are three types of RFID tags basically:

First are the active tags: these types of tags are directly connected to the battery and are used to transmit signals sporadically, and they have very strong signals as well as

S. R. Singh
Department of Mathematics, C.C.S. University, Meerut, India

R. B. Singh · A. Rastogi (✉)
Department of Statistics, D. N. College, Meerut, India
e-mail: akshikastats@gmail.com

S. Singhal
Department of Statistics, Vardhaman College, Bijnor, India

M. Sahni et al. (eds.), *Novel Developments in Futuristic AI-based Technologies*, Algorithms for Intelligent Systems, https://doi.org/10.1007/978-981-99-3076-0_3

a wide range, so they are used mostly in location tracking. Second is the passive tags which are very much used in the commercial field. These tags are not active unless they receive a signal from the reader, and transmit the signal back as information. Third are the Semi-active tags; although they are connected to the battery, they do not transmit signals sporadically.

In this paper, the role of passive tags has been discussed which are mostly used in supermarkets. They do not have a very wide range, so they work in near-field communication (NFC).

2 Literature Review

In recent years, RFID has been a center of attraction for researchers and scientists. Many scientists are still working in that a direction. Various costs are related to the RFID models, and further these costs are compared with EOQ models to show the merits over it [7]. JIT wireless manufacturing framework was the one in which objects are traceable with the help of RFID as it is a real-time visibility [18]. A simulation model was proposed [6] to study the effect of the collision of two or more tags which created the situation of false alarm. Further, they explained how reading time is affected by the magnitude of the population. Similar kinds of wireless techniques are used in other sectors also; [10] showed the standards of wireless techniques in the agriculture sector. Through a case study, [3] found that RFID technology is supportive of the library also. Turkey library area is chosen as a sample to investigate. In the inventory field, a producer must deal with an amount of unknown uncertainty. Much research has been done on this kind of situation, and the model is demonstrated in the fuzzy sense. A model covered [11] the practicality of inventory models. A work was proposed [2] where the role of RFID is explained in the day-to-day shopping of a person and how it would be able to save time and take care of the economy. In the same year, [8] examined that RFID technology is fruitful to keep the inventory in control as keeping inventory is an essential job for the growth of a business. A business firm must take the consideration of various items. [12] considered an inventory system with a random probability distribution for decomposed items with volume flexibility, and further demand is considered to be exponential. Later [13] explained an inventory model that depend on time and stock with finite time horizon [17]. Investigates that how the energy can be improved and how passive RFID tags increase efficiency during operational activities. In the same year, [9] proposed in their work how RFID technology is used in billing software overcome the technology of barcode and the information will be automatically sent to the host system, and it shows the real-time visibility. [4] explained that RFID technique has many merits over traditional systems; it is proven to be supportive in every dimension. A mathematical model was developed [14] for decaying items with multiple markets with volume flexibility. RFID is trending in Construction work also [16]. For example, the technology is used for safety, quality of products, and to improve the efficiency of construction work. A paper was reviewed [1] about the security and privacy issues solved by the RFID technique. Many researchers have tried to find ways to tackle the pandemic; [15] explained similar techniques to JIT which is very economical as compared to traditional techniques. In the same year, RFID was accepted as a new method to achieve sustainability in their model and was proved to be a decision-making technique also [5].

In this model, it is explained how the cost cut can be done once RFID is implemented in the system. Through a numerical example, it is explained that it benefits the EOQ system in terms of cost.

3 Methodology

To understand the concept in depth, the case study is considered to be very effective. The same study has been done in this paper to understand the concept of RFID in detail. For many years, many commercial organizations are using RFID technology in their inventory management. So, for this, many short interviews with several managers and sales executives have been conducted. Some telephonic calls are also being made. Many commercial stores like Big Bazaar, Easy Day, Vishal Mega Mart, etc. have been being visited during the study that has already used this technology. In some stores, it is found that they are using this technology for the past two years and some are using it for a long time. The method is based on observation and interviews mainly.

The sample for the stores is taken at Radom. Random Sampling has been done only of the stores which are using RFID technology. Many Shopping stores are taken randomly as a sample. Questions have been asked in Chromatic or descriptive form.

4 Call for RFID in Shopping Style During Pandemic

The shopping style of customers during the pandemic has changed drastically. Customers are more aware now of their surroundings. Health and fitness have become a priority now. Maintaining social distancing has become a challenge for a producer also as they have to take care of customers' demands.

Social distancing, maintaining inventory, and fulfilling customers' needs are a big challenge during the pandemic. During pandemics as physical contact is not there. A producer has to take care of both terms. Shopping habits have changed over the night during this tough time and accordingly, supermarkets have changed their working style also. Mostly, the offline orders have changed to online orders. Real-time visibility is required every time so that the producer can take care of every need of a customer. During a pandemic, RFID technology plays an important role. RFID is nearly a new technology around us and now it is emerging in the commercial field day by day. If we talk about inventory management, so in that area not much physical touch is required which proved to be a boon in the time of the pandemic.

On the other hand, it overcomes the demerits of barcodes. Many of the supermarkets and shopping stores were using the technique of barcode. But in the barcode system, physical touch is required as the reader has to scan the tag of the product every time.

But in the RFID technique, the tags transmit the information to the system, and the inventory is updated continuously in the system of shopping centers. In the supermarkets or stores when they are getting orders both online and offline, maintaining the inventory is a must. A producer got the status of inventory through RFID tags, and they can easily fulfill the demand of customers.

5 Case Description

Many technicalities are required when you must deal with a mass. This case study has been done to understand how RFID technology has supported the system and how a manager of the supermarket/store keeps a record of inventory.

It is no doubt that because RFID required less physical contact, it is substantiated during the vital time of the pandemic. Its real-time visibility approach has helped a producer/manager to keep track of inventory. During the pandemic, both the online and offline demands are coming. Shoppers prefer not to go to the store and order online. So, by keeping the RFID tags on the product, it would be easy to keep the status of the inventory updated without very much physical contact. Several stores have been visited during the case study to understand how the products are received by the store, how they manage the things, and what is the role of RFID.

There are certain steps that are taken one by one.

All the products which are kept in the store have passive RFID tags already before they entered the store. These tags have some specific code hitch that can give the information of city, store, product, price, etc.

Every store has varieties of products in their store. They have food items, dairy products, kitchenware, etc. So, it is necessary to have tags on all the products, once all products are unloaded (Fig. 1).

First, the products are cross-checked before they are taken into the store. After that, all the products are entered in the store computer through ERP and SAP software. These days, mostly stores are using SAP, a statistical software to keep a record of the inventory along with their code. Each store is provided a plan to keep the inventory in order. It changes from time to time; it is mostly seasonable.

If a customer is picking something offline, they would go to the billing counter and an automatic bill wii be generated through replacement billing software. As the products are having RFID tags, the inventory is updated in the system through SAP. Every store must keep a minimum BASE Quantity (MBQ) of every product in their store. It is according to the requirement of a product and the warehouse itself decides about the MBQ of every product. So as the products are sold through the passive

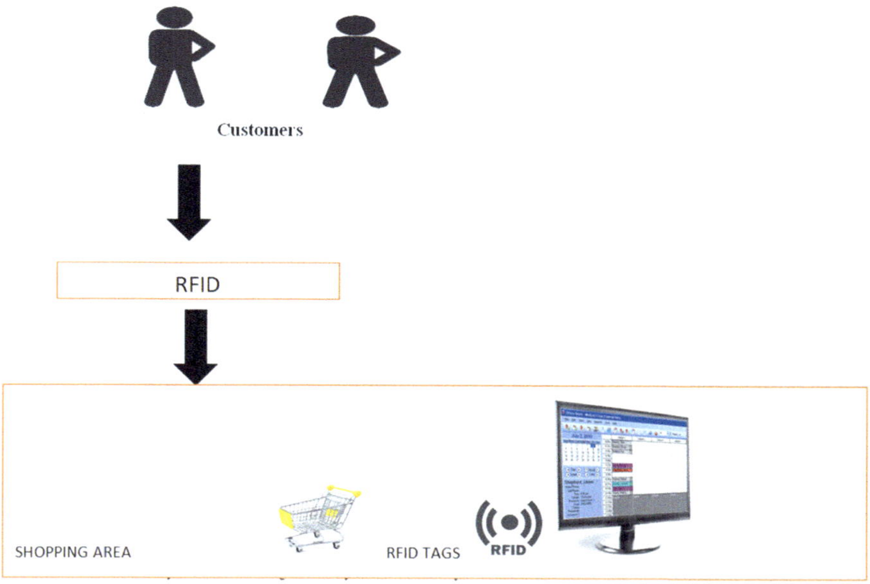

Fig. 1 Mechanism of RFID

tags, they are automatically updated in the system; the demand will automatically go to the warehouse and the store will get the product requirement within 4–5 days or within a week to maintain MBQ of every product.

In the same manner in the case of an online scenario also, the work of passive RFID tags will remain the same and it will update the system to maintain the inventory and send the demand to their respective warehouses to balance MBQ.

6 Notations

O.C	Ordering cost
S.C	Shortage Cost
Cc	Carrying cost
F	Fixed Cost
v	Variable Cost
EOQ_{RFID}	EOQ model after RFID implementation
v_{RFID}	Variable cost after RFID implementation
R	Tag Reader
F_{line}	Cost of Ethernet connection

(continued)

(continued)

O.C	Ordering cost
F_{switch}	Cost of Ethernet Switches
k	No. of ports
v_{eth}	Maintenance cost for rewiring of ethernet
TIC	Total annual inventory cost
TIC_{RFID}	Total annual inventory cost of RFID system
B_S	Reduction cost in shoplifting due to RFID
B_{po}	Reduction cost of checkout due to RFID

7 Mathematical Modeling

RFID is a technique which has multiple dimensions and has a good influence on the system. If this technique is introduced in the system, then it will have benefits in terms of cost and time, and on the other side, it fulfills the SOP of the pandemic also. For this model, we explained that the fixed cost for RFID implementation is more beneficial as compared to the ordinary EOQ model. Carrying cost consists of fixed cost and variable cost and the overall cost included in the implementation of RFID is validated to be less comparatively.

For an ordinary EOQ model:

T.C (Total Cost) = Ordering cost + Shortage Cost + Carrying Cost

$$\text{T.C.} = \text{O.C} + \text{S.C} + \text{Cc} \tag{1}$$

Carrying cost includes Fixed and Variable costs also:

$$T.C. = O.C + S.C + (F + v) \tag{2}$$

If for a particular store F is the fixed cost and v is the variable cost, so the total carrying cost will be F + v m, where m is the total number of replacements in a year.

And EOQ = $\sqrt{v}$

And after RFID, it will be

$$EOQ_{RFID} = EOQ_{RFID}\sqrt{v_{RFID}/v} \tag{3}$$

Then,

$$T.C = O.C + S.C + (F_{RFID} + v_{RFID}) \tag{4}$$

where F_{RFID} is the fixed cost after RFID installation, and it is the initial cost.

It is the cost of certain things that are used for complete installation like wirings and switches. Let R be the number of tag readers. So for the desired model, the formula will be (F + v) R.

Now,

$$F_{RFID} = R\,F_{line} + \frac{R}{k}F_{switch} + R\,v_{eth} \tag{5}$$

v_{eth} is the maintenance cost for rewiring of ethernet and power line w nodes:

$$v_{RFID} = R\,F_{tag} + R\,v_{wifi} \tag{6}$$

$$T.C. = O.C + S.C + R\,F_{line} + \frac{R}{k}F_{switch} + R\,v_{eth} + R\,F_{tag} \tag{7}$$

Further, the model total carrying cost is the cost of the shelf space occupied by the item.

Let TIC_j be the total annual inventory cost for the jth item and

$$TIC = \sum_j TIC_j \tag{8}$$

$$TIC_{RFID} = TIC_j\sqrt{v_{RFID}\,/\,v}\,\text{forthe}\,j\,\text{item} \tag{9}$$

$$TIC_{RFID} = \sum_j TIC_j\sqrt{v_{RFID}/\,v} \tag{10}$$

$$TIC_{RFID} = TIC\sqrt{v_{RFID}/\,v}$$

For this system, Total System Cost is

$$\mathrm{TSC} = \mathrm{F} + \mathrm{TIC} \tag{11}$$

$$TSC_{RFID} = F_{RFID} + TIC\sqrt{v_{RFID}/\,v} \tag{2}$$

The two major benefits of RFID are also included

- Overcome with loss of Shoplifting
- Reduction in checkout cost.

Let B_S be the benefits due to the reduction in shoplifting from RFID and B_{po} be the reduction of checkout due to RFID.

Taking both costs into considerations,

$$F_{RFID} + TIC\sqrt{v_{RFID}/vB_sB_{po}} + \mathrm{F} + \mathrm{TIC} \tag{13}$$

The RFID system is said to be profitable if

$$F_{RFID} < F \cdot B_sB_{po} + TIC\left(1 - \sqrt{\frac{v_{RFID}}{v}}\right) \tag{14}$$

$$F_{RFID} < EOQ \quad \left[EOQ = F \cdot B_sB_{po} + TIC\left(1 - \sqrt{\frac{v_{RFID}}{v}}\right)\right] \tag{15}$$

This can be further illustrated by the numerical example.

8 Numerical Exemplification

Suppose there are three retail stores. The fixed cost and the variable cost for different retail stores are given in the below table. If these retail stores have RFID amplification, there will be several costs involved in the company, some are fixed and some are variable. The fixed cost is the cost included for RFID replenishment.

As there will be RFID amplifications, there will be some benefits of this replenishment like there will be savings due to checkout terminals and shoplifting cases. Let the annual cost be 980,000, 1,200,000, and 1,800,000 respectively for the Retail stores A, B, and C.

For Retail store A $\mathrm{F_{RFID}} = 750000$ and $\mathrm{v_{RFID}} = 0.1$, $B_S = 365$, $B_{po} = 24000$, F = 312,000, v = 0.75.

For Retail store B, $\mathrm{F_{RFID}} = 900000$ and $\mathrm{v_{RFID}} = 0.1$, $B_S = 4500$, $B_{po} = 30000$, F = 410,000, v = 0.75.

For Retail store C, $\mathrm{F_{RFID}} = 1500000$ and $\mathrm{v_{RFID}} = 0.1$, $B_S = 7000$, $B_{po} = 50000$, $F = 700{,}000$, $v = 1.25$.

Calculating EOQ for each retail store and putting the values in Eq. (15), we compare the fixed values for proving the benefits of RFID (Table 1).

As we can interpret from the table Fixed cost by implementing RFID is more economical and in each store, there is a saving due to these implications (Fig. 2).

From the above diagram, we can interpret that there is a huge benefit of RFID implications on the EOQ model (Fig. 3).

Table 1 Comparison of EOQ model with and without RFID

Retail Stores	$\mathrm{F_{RFID}}$	$\mathrm{v_{RFID}}$	$\mathrm{B_s}$	$\mathrm{B_{po}}$	F	v	TIC	EOQ	Saving
A	750,000	0.1	3650	24,000	312,000	0.75	980,000	961,805	211,805
B	900,000	0.1	4500	30,000	410,000	0.75	1,200,000	556,100	343,900
C	1,500,000	0.15	7000	50,000	700,000	1.25	1,800,000	2,431,000	931,000

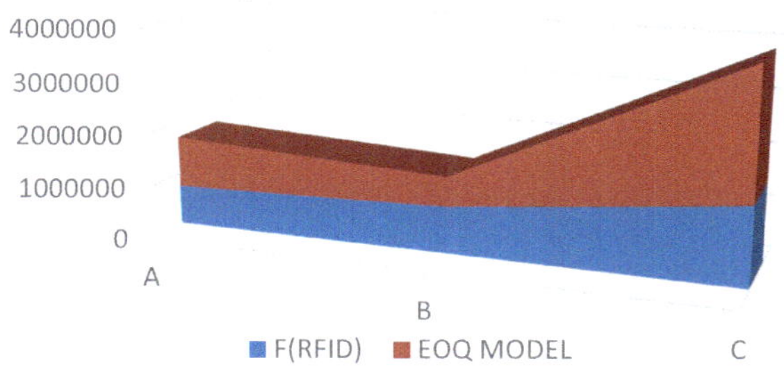

Fig. 2 Fixed cost of RFID model versus EOQ model cost

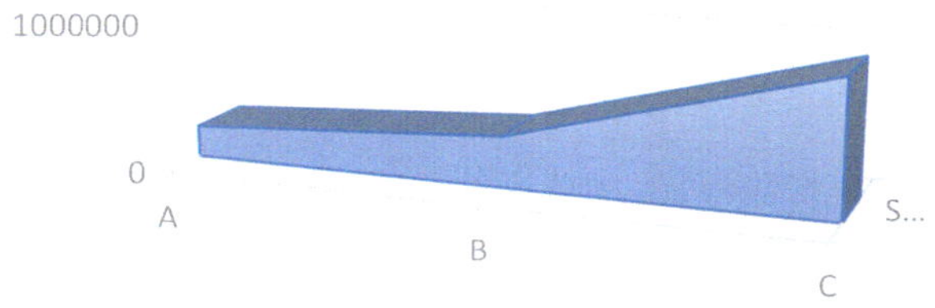

Fig. 3 Graphical representation of saving cost per model

The above graph reflects that there is a saving in each model by RFID implications on the EOQ model.

9 Conclusion

Incorporating the RFID technique is sustainable as it saves the energy of a manager, saves time, and cuts the cost also. In the traditional EOQ model, there is so much involvement of manpower and energy in keeping control of the commerce and trading sectors. Presently, RFID is very serviceable in the commercial field, which needs several orchestrations. In this paper, we have deep knowledge of these arrangements through a case study. These implementations have many benefits over the regular methods to keep a watch on the inventory and at the same time, it fulfills the protocols of a pandemic. There are several costs involved in these procedures. Through a numerical iteration, it is found that the fixed cost involved in RFID implementations is less than the traditional EOQ model.

As technology is updated day by day, the commercial department is updated so far, and keeping track of inventory is not so complicated these days. The RFID system has a future in many other departments also like medical, defense, education, etc. It fulfills the protocol of social distancing as well so this system can be further used as a safeguard.

References

1. Ausecha CM, Rosero J, Gonzalez GR (2021) RFID applications and security review. Computation 9(6):69
2. Ali Z, Sonkusare R (2014) RFID based smart shopping: an overview. In: International conference on advances in communication and computing technologies (ICACACT 2), pp 1–3
3. Aydın K, Yıldırım S (2012) Case study about RFID system in library services. Int J Synerg Res 1(2):91–102
4. Delgado MV, Burrull F, Mariño P (2017) Radio frequency identification. In: Chapter6, case study: installing RFID systems in supermarkets, pp 121–139
5. Kluczek A, Gladysz B, Ejsmont K (2021) Application of lifecycle measures for an integrated method of environmental sustainability assessment of radio frequency identification and wireless sensor networks. Energies 14(10):1–22
6. Kamarthi S, Benneyan J (2008) Analysis of communication between RFID tags and their reader using statistical simulation model. Int J Comput Appl Technol 32(21):28–134
7. Madni A, Boppana R, Chalasani S, Sounderpa J (2007) Models for cost-benefit analysis of rfid implementations in retail. IEEE Syst J 1(2):105–114
8. Poojary A, Kumar RS (2014) RFID application to improve inventory management. IJMBS 4(4):1000–1003
9. Indra Reddy YC, Keerthana N, KavyaShree S, Deepthi M, Shiva Prasad KM (2017) Automated billing system in super markets using RFID technology. Int J Sci Res (IJSR) 6(6):1680–1681
10. Ruiz-Garcia L, Lunadei L, Barreiro P, Robla I (2009) A review of wireless sensor technologies and applications in agriculture and food industry: state of the art and current trends. Sensors 9(6):4728–4750
11. Singhal S, Singh SR (2013) Volume flexible multi-items inventory system with imprecise environment. Int J Ind Eng Comput 4(4):457–468
12. Singhal S, Singh SR (2014) A production remanufacturing system for probabilistic decaying items with volume flexible environment. In: Proceedings of 3rd international conference on recent trends in engineering and technology (Elsevier), pp 431–443
13. Singhal S, Singh SR (2015) Modeling of an inventory system with multivariate demand under volume flexibilityand learning. Uncertain Supply Chain Manag 3(2):147–215
14. Singhal S, Singh SR (2018) Supply chain system for time and quality dependent decaying items with multiple market demand and volume flexibility. Int J Oper Res 31(2):245–261
15. Singh SR, Rastogi A, Singhal S (2021) JIT: the best approach after lockdown in country. Int J Serv Oper Inform 11(1):75–86
16. Spisakova M, Kozlovska M (2020) Application of RFID technology on construction site–case study. In: VIII international scientific siberian transport forum, pp 29–36
17. Tsiropoulou E, Baras J, Papavassiliou S, Sinha S (2017) RFID-based smart parking management system. Cyber-Phys Syst 3(1–4):22–41
18. Zhang Y, Jiang P, Huan G (2008) RFID-based smart Kanbans for Just-In-Time manufacturing. Int J Mater Prod Technol (IJMPT) 33(1–2):170–184

Chapter 4
Enhancing Agriculture Supply Chain Management Using Blockchain

Sandra Samuel, R. Maheswari, and N. Mohana

1 Introduction

A product's life cycle consists of many stages such as Produced (Production Stage), Cultivated (Cultivation Stage), Packed (Packing Stage), Labelled (Labelling Stage), Stored (Storing Stage), Shipped (Shipping Stage) and Sold (Selling Stage). The stage at which the product is determined by the respective supplier who is currently in hold of the product. A product transition happens between two stages, if the two stakes are linked. A Smart Contract is a digital contract or a ledger which validates the transactions between the stages. A digital Smart Contract is a ledger or a programmed document that automates and governs the relationship between stakes involved in the transition stages of a product.

1.1 Background Statement

The food sector faces various constraints which hinder the distribution of culturally preferred fresh food. With a lack of suitable infrastructure, the mechanical supply chain management is inefficient, involves middlemen and has no transparency of transactions. There is no adequate quality control and testing. The taxation costs of inventory and packaging are far more expensive than what it would actually cost. The growing concerns have enhanced the need for a traceable system that is transparent, trustable, secure and efficient.

S. Samuel · R. Maheswari
School of Computer Science Engineering, Vellore Institute of Technology, Chennai, India
e-mail: maheswari.r@vit.ac.in

N. Mohana (✉)
School of Advanced Sciences, Vellore Institute of Technology, Chennai, India
e-mail: mohana.n@vit.ac.in

M. Sahni et al. (eds.), *Novel Developments in Futuristic AI-based Technologies*,
Algorithms for Intelligent Systems, https://doi.org/10.1007/978-981-99-3076-0_4

1.2 Motivation

Product safety is an important aspect that requires constant monitoring of agricultural products and logistics management. Products traded across several countries require accurate tracking information to validate every transaction presented by all stakeholders involved in its life cycle. This involves connection, communication, authentication, management and transfer of crucial data, its origin and other information exchanged. This complex ecosystem has a long chain of stakeholders who play various roles in a product's life cycle from farm to product. The proposed system of smart contracts removes the need for a trustable source of central authority and other intermediaries. All information from end to end is stored in an encrypted immutable ledger of a Blockchain, and the transactions are governed over by a smart contract. Figure 1 shows the results from the PwC Blockchain survey of the top three challenges depicting the difficulty levels in solving the problem.

2 Literature Review

The two main purposes of updating a product state: To Validate the transition and transaction and maintain the state of local data. The transactions in the Smart Supply contract update the product state in supply chain management [1]. The basic structure of the Blockchain is shown in Fig. 2.

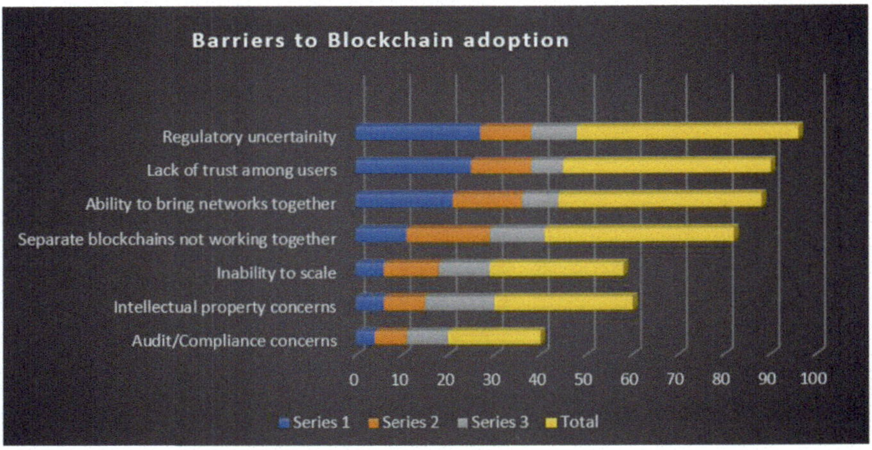

Fig. 1 Results from PwC blockchain survey of top three challenges

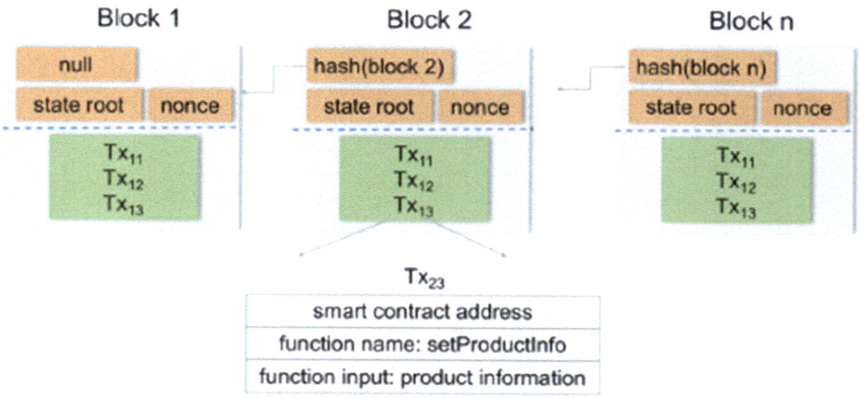

Fig. 2 Structure of the blockchain

Any transaction consists of the following fields:

- *Address*: It throws reference to the part of code in the Blockchain that defines a smart contract. Details on the final state of the products in the Smart Contract are updated here.
- *Function call*: It points to the change in the state of internal functions of the Smart contract.
- *Function Parameters*: They point to the uniquely identified parameters of the smart contract.

2.1 Limitations of Ethereum Smart Contracts

The validation of a transaction cannot be changed in Ethereum. Ethereum's validation process is application-agnostic [1]. Every transaction is validated before being added to blocks. While transactions are processed, the miner does not tamper with the contents of the transaction. Validated product information is kept in the local data structures of the miner that can be queried and retrieved.

Limitations of Stack memory: A smart contract function in Ethereum can contain only 16 variables, whereas a typical system can contain hundreds. A smart contract can use hierarchical variables to bypass stack memory limitations.

Blockchain technology paves the way for two parties to verify a distributed ledger through verifiable and immutable means [2]. Transactions are governed by public/private cryptography indicated in Fig. 3. All valid transactions are updated on the chain in terms of new blocks. An efficient bridge was proposed [3] to gap between the literature and conceptual application of blockchain in agriculture supply chains by showing how Ethereum smart contracts offer security, decentralized trace and track solutions with adequate architectures and traceability.

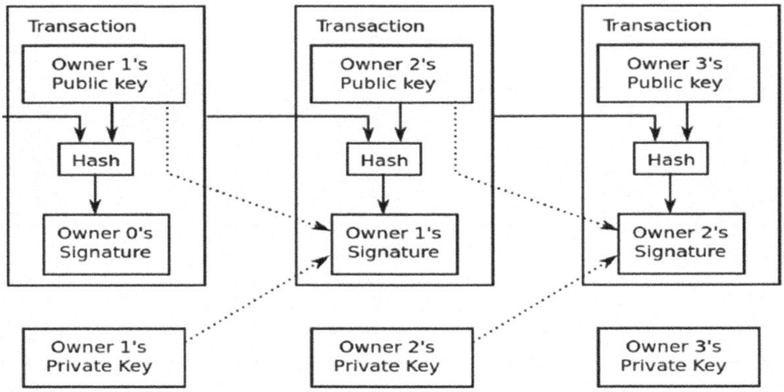

Fig. 3 Blockchain principle

2.2 *Traceability*

Continuous monitoring is essential compliance to further the quality of products and also achieve transparency between the multiple stakeholders. The quality conditions can be monitored using IoT-enabled packaging equipment which allows traceability of produce from precise physical locations using global location identifiers or by using geo-tagging equipment with GPS sensors. A stakeholder also has the power to transact and track its fraudulent data by recording its footprint. To overcome this type of fraud, additional functionalities to invalidate shipment or impose penalties on fraudulent stakeholders have been added. This ensures accuracy, undisputed traceability and auditability. A storage mechanism of bitcoin data was proposed [4] to trace counterfeit medicines. Every member in the link will share their corresponding public key along with the hash value of their previous transaction pertaining to the encrypted Quick Response (QR) code provided by the manufacturer. The transaction is made secure and tamperproof. The structure does verification using the sender cryptographic signature and a unique QR code. A relative analysis of the smart contracts in view of its market demand holding its relevant supply metrics is presented [5, 6]. Their results indicate that uniform k-Double Auction (k-DA) has the highest percentage of trade and signed contracts. Budget balance in the market ensures the prevention of subsidizing or trade gains between buyers and sellers.

2.3 *Types of Blockchain*

Figure 4 represents the existence of various types of blockchains.

- Public: It allows any individual to access, submit and participate in consensus procedures. The blockchain is open and transparent though the privacy and

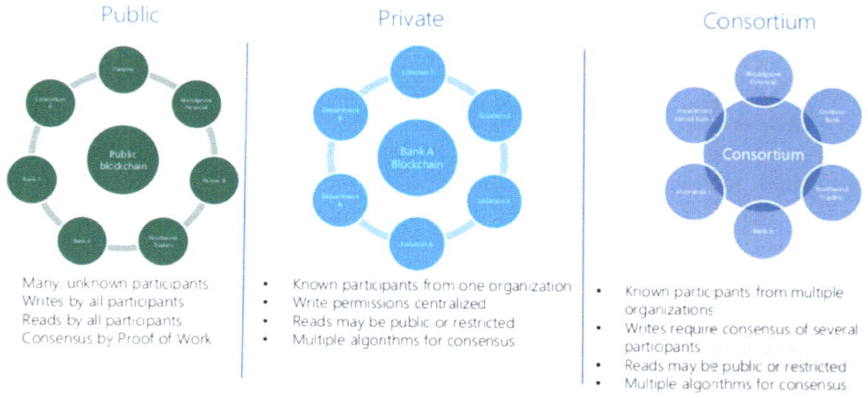

Fig. 4 Types of blockchain

anonymity of the nodes are protected. Bitcoin, Ethereum and Litecoin have adopted public blockchain.
- Private: Access permission is limited and requires a trustworthy agent for consensus to be reached. It adopts a consensus object that is different from a public blockchain.
- Consortium: This kind is partially decentralized where the blockchain refers.

2.4 Oracle

An oracle is a key link between Blockchain and the physical world [7–11]. Blockchain, by design, does not have the provision to access data from the outside. An oracle plays the integral role of a trusted intermediary and facilitates data to be fed into the blockchain ecosystem. Oracles can be classified under three categories:

- Hardware-based (sensors, IoT and weather stations).
- Software-based (expiration date, New York Stock Exchange and computational output).
- Consensus-based.

2.5 Markov Process

Markov process, a basic mathematical tool, is used to estimate the performance of a blockchain system [12]. It has been developed with a broad framework holding the block structure of Markov processes and provides an efficient methodology for manipulating the average transaction endorsement period for the specified transaction. Moreover, the given performance measures are the sufficient key to improve the performance of blockchain. The generalization of the 2-stage queueing model batch-service discussed the varying range of all Poisson to transaction of MAP with its arrivals and also includes the exponential to *Phase Type Service* epochs.

To estimate the performance, the system introduced a 2-stage design, namely *Service In Random Order* and queueing system batch service, along with all relevant details such as transaction occurrences of MAP and PH service times with 2-stages. It also measures some key performances, including the mean transaction for stationary items in the waiting room of the queueing system, the mean stationary transactions in the block, and the mean transaction confirmation time of any transaction. Moreover, the given performance measures are the sufficient key to improve the performance of blockchain. The generalization of the 2-stage queueing model batch-service discussed the varying range of all Poisson to transaction of MAP with its arrivals and also includes the exponential to *Phase Type Service* epochs.

2.6 Hypergraphs

Hypergraphs are used to simulate computer networks, biological networks, process scheduling, data structures and various other systems. The intelligent devices connected to the Internet of Things (IoT) are characterised by lightweight systems that consume very little energy and memory capacity. This characteristic makes it tough to incorporate blockchain into these devices. To overcome this memory limitation, a blockchain model has been initiated based on hypergraphs [16], which aim to diminish storage consumption to resolve overhead security threats.

The study of hypergraphs has been done [13] to discover the network's storage of records by considering each home as a vertex in the graph. Also, the response strategy, security risks and their working mechanism have been discussed. The proposed model of blockchain using hypergraph can be deployed to facilitate privacy protection requirements and maintain the security of smart homes. Storage capacity has been increased from the existing blockchain model by employing the concept of hypergraphs. Also, the mean storage in every vertex has been increased pertaining to its rank available in the system of *hyperedge* which is disjoint from its graph *degree*.

A method has been presented for improving the execution of blockchain by extracting the blockchain network in a hypergraph [14]. The safety and security risk for reasonable stage and strategies of response through network settings have been discussed as a security problem for this model.

2.7 Queueing Model

Queueing theory is a very useful mathematical model in the blockchain system for its performance exploration and to ensure its increased performance through optimization. System performance evaluation and queueing theory of blockchain systems have been discussed [15]. Also, the design of Markovian *Batch Service Queueing System* has been given with a couple of distinct service stages, which are effective for the building of a new blockchain and the mining process in the miners' pool. Key performance measures, namely the average transaction-confirmation time along with the number of transactions that occurred on average in a block and also the number of transactions that happened on average in the queue have been expressed by matrix-geometric solution.

A discrete-time queueing model has been developed [16] to permit the assessment of significant performance blockchain indicators along with its distributed ledger. It predicts the queue size and average waiting time of the transaction for its allocation, which depend on different input factors. It performs a study of the very first input parameter to determine the result of its distinct parameters on the behaviour of the system. The validity of the proposed model has been investigated in a controlled environment that depends on its private Ethereum based blockchain. It also verifies the extrapolations of the suggested system model and embraces the concept of *Proof of Authority* included in the mechanism of consensus.

3 System Design

Smart Contracts will receive function calls as transactions. They also send a trigger event to all the participating entities every time a node is mined. This allows constant monitoring to track and be alerted when violations occur. Every participant in the blockchain is expected to have a requirement: An Ethereum account with a unique Address (EA), which contains public and private keys which are cryptographic digital signatures that serve to validate the authenticity and integrity of the data in each transaction and association with its respective EA.

3.1 The Consensus Algorithm

For the proper functioning of a blockchain, it requires a system of consensus. This consensus is the key mechanism responsible for all the validations of the blocks, added or appended as per the agreement or contract signed. Distributed systems that use decentralized blockchains do not rely on a central algorithm and depend on distributed nodes to agree or consent to the validity of transactions. The system works on the basis of a consensus algorithm which is the integral mechanism involved in the creation of new blocks appending them to the blockchain. The popular consensus algorithms include Proof-of-Concept (PoC), Proof-of-Work (PoW), Proof-of-Authority (PoA) and Proof-of-Stake (PoS). Figure 5 gives the insight of comparison of various Consensus Algorithms.

characteristics	consensus algorithms				
	PoW	*PoS*	*DPoS*	*PBFT*	*RAFT*
Byzantine fault tolerance	50%	50%	50%	33%	N/A
crash fault tolerance	50%	50%	50%	33%	50%
verification speed	>100s	<100s	<100s	<10s	<10s
throughput(TPS)	<100	<1000	<1000	<2000	>10k
scalability	strong	strong	strong	weak	weak

Fig. 5 Comparison of consensus algorithms

3.2 *Byzantine Fault Tolerance (BFT)*

The dependability of distributed computing nodes is defined by BFT, where two nodes can safely communicate, well aware of having to display the same data. This is one of the most complicated challenges that Blockchain technology addresses. Active systems must be able to adopt strategies that save them from total system failure. Without BFT, a blockchain is unreliable and prone to malfunction instances. Hyperledger, Stellar, Dispatch and Ripple are some works which have successfully implemented the BFT algorithm.

3.3 *Proof-Of-Concept (PoC)*

The PoC pilot demonstrates that a certificate-based blockchain is feasible with a permission-based ledger and a smart contract. PoC shows how different stakeholders reside on the same layer of information on the basis of the contract providing tamper-proof validity, which potentially increases the values of certificates. The limitation of PoC is that the amount and quantity of data and information shared are of very small quantities.

3.4 *Proof-Of-Work (PoW)*

A PoW consensus algorithm is one in which an intense computational puzzle has to be solved for a node to acquire the right to publish a block. The computation results are verified, and it aids other nodes to validate and update the blockchain. The successful solving of the puzzle is termed as 'Mining'. PoW-based blockchain defend Sybil attacks by directing miners to solve expensive computational tasks or problems. Mining is basically a method that involves finding a random number that will add a certain required number of zeros before the sequence of the SHA-256 hash at the block header. The complexity is determined by the number of zeros, the more zeros, the more the complexity.

3.5 *Proof-of-Stake (PoS)*

Other block verification methods such as proof-of-stake are subsequent spin-offs of proof-of-work where a stake takes the place of the energy and computational overheads. In PoS, it is assumed that those with the most participation are less likely to attack it. The drawback of PoS lies in the fact that a node with rich digital assets can earn the right to create new nodes.

3.6 *Stakeholders*

The factor which determines if a food product could be produced and whether it is consumed in a maintainable manner without compromising its cost lies in the sustainability of its supply chain which involves stakeholders reaching outside organizational boundaries for the well-being and upliftment of current technology along its regulation in market and other socio-economic deliberations. Effective collaborations involve tackling food loss and food waste for a start. Corporates and parties play crucial roles in food chains. The existence of adaptive management and its related governance to advance the state of the economy and nutritional security require considerable enhancement in the global system for the integral functioning of the stakes involved. The key Stakeholders involved in the Agriculture supply chain are depicted in Fig. 6.

The primary apprehensions of any stakeholders involved are summarized as follows:

- How much of the data in the blockchain is reliable based on the data's origin;
- Environmental impact of mining;
- Dealing with new blocks added to the chain;
- Validating transactions in the smart contracts;
- The efficiency of the system in identifying and avoiding scam activities in the supply chain of food.

These apprehensions could be mostly resolved with the advancement in technologies such as the use of authentic data that is machine-generated data or by developing the principles for blockchain and its corresponding smart contracts.

The main purpose of adopting Blockchain technology is because of its cryptic data security. While data tampering can be prevented, the Blockchain cannot prevent or keep stakeholders from adding false or error-prone data or hiding data from the

Fig. 6 Key stakeholders involved in agriculture supply chain

system. Although, when abundant data is available, cross-checking and immutability of records might trace out frauds, this may also reduce the prospect of wrong data from entering the network. System architecture indicating various modules/components and their interaction is shown in Fig. 7.

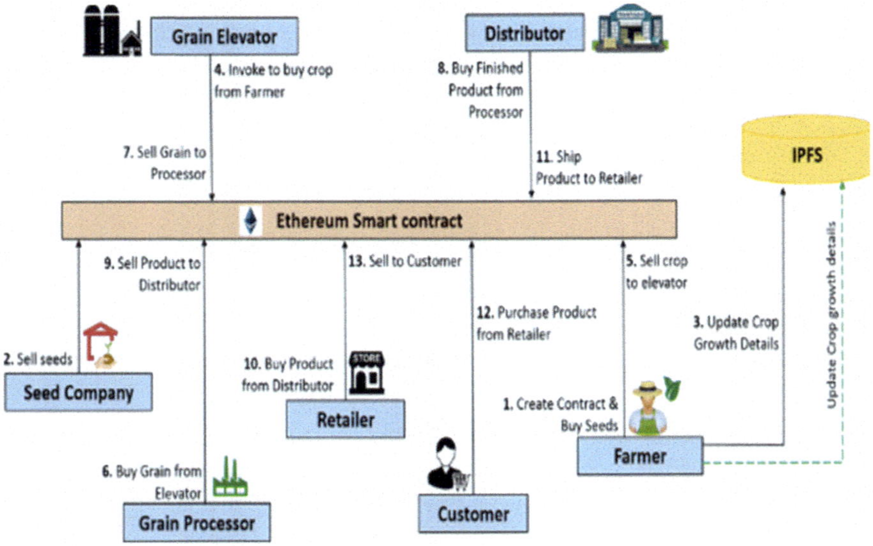

Fig. 7 Architecture of the supply chain

4 Implementation of System with Its Methodology

To build a decentralized application, the first step is to generate a peer-to-peer network system with all nodes that are to take part in the Blockchain. Each and every node in the chain of the network receives two keys: a public key and a private key. The keys bring about encryption and decryption. The public key allows the users to encrypt messages sent to the node. It works like a unique address. The private key allows the node to read the received information, and to approve and sign transactions in the Blockchain. To decrypt, the user must have a correct private key for the corresponding public key.

4.1 Sequence Diagram

A transaction is carried out with a node signing it and transmitting to it peers. This enables authentication and guarantees integrity as it prevents any form of invalid decryption to take place. The peers broadcast the signed transaction after verifying it, which again is retransmitted to the other nodes throughout the network. The valid transactions are assembled into a time-stamped block by special nodes known as 'Miners'. How and what data is inserted are chosen on the basis of a consensus algorithm. The sequence diagram for the proposed system is represented in Fig. 8.

The blocks are transmitted or shared back into the network where the system verifies its authenticity with its references to its previous blocks based on the hash address. If they validate negative, the respective block is not accepted. If the premonitions are satisfied, the block is added to the existing length of the chain in the network.

4.2 Remix

Remix is a tool that is open-sourced which enables writing Solidity contracts straight from the browser. When a compilation succeeds, a JSON file for the compiled contract is created by Remix. This JSON file contains the compilation's artefacts and metadata. The JSON file contains links to the bytecode, libraries, gas estimation, method identifiers and the ABI.

4.3 Ganache

Ganache is a virtual blockchain which sets up 10 Ethereum addresses by default, completes with private keys and pre-loads them with 100 simulated ether balances

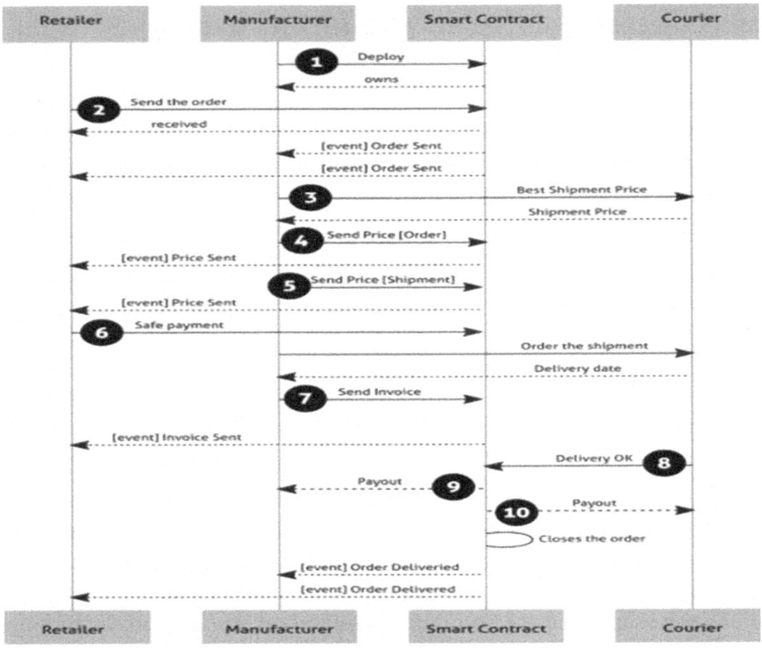

Fig. 8 Sequence diagram

each. Ganache doesn't involve mining, but it simply confirms transactions coming its way which makes iterative development possible. With the virtual node running, we can connect it to a wallet, MyEtherWallet. The Ethereum Smart Contract deployment architecture is given in Fig. 9.

5 Results and Discussions

This system has proposed a blockchain model to produce a smart contract in the Agriculture Supply Chain. This work validates the correct execution of softwares to present that PoC can support use cases of Ethereum and Solidity. The transactions in the system are transparent, which provides chances to avoid fraudulence during the overall process. The stakeholder is aware of the provenance and timestamp of the product in the supply chain. Tracking is made possible by the system.

Detailed procedural results for each logical component of the proposed system with screenshots are represented in the below section. Figure 10 gives insight into the create custom node with RPC server URL and PORT from Ganache.

Figure 11 ensures the deploy Contract with extracted byte code of solidity code from Remix.

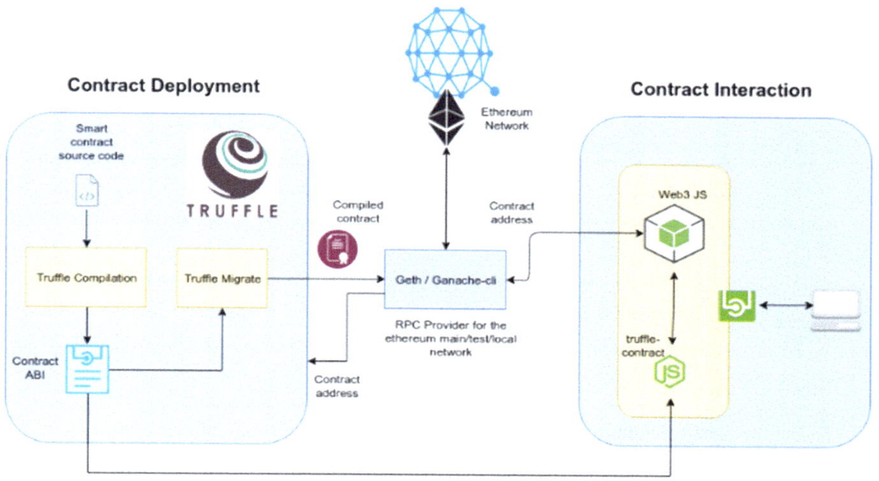

Fig. 9 Ethereum smart contract deployment architecture

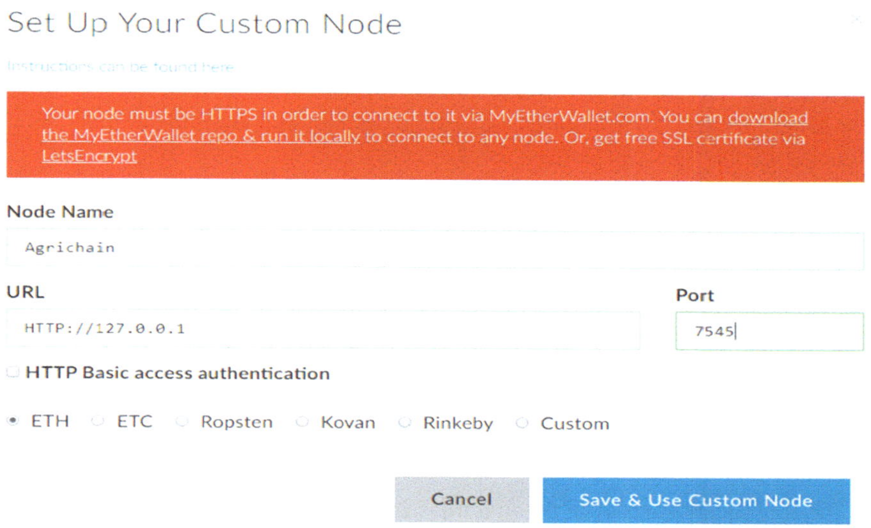

Fig. 10 Create custom node with RPC server URL and PORT from ganache

Decrypt and access Wallet using a private key from one of the 10 addresses available in Ganache is represented in Fig. 12.

The sign and deploy contract as the first transaction along the byte code generated with a raw transaction is illustrated in Fig. 13.

The success of its first transaction and byte code is demonstrated in Fig. 14.

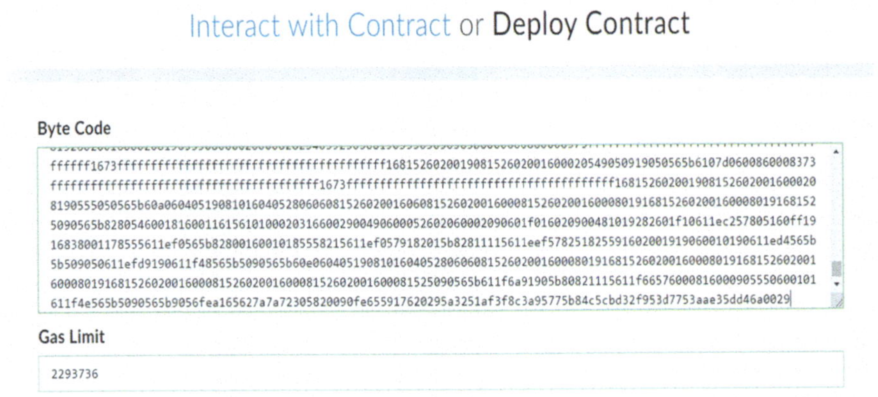

Fig. 11 Deploy contract with extracted byte code of solidity code from remix

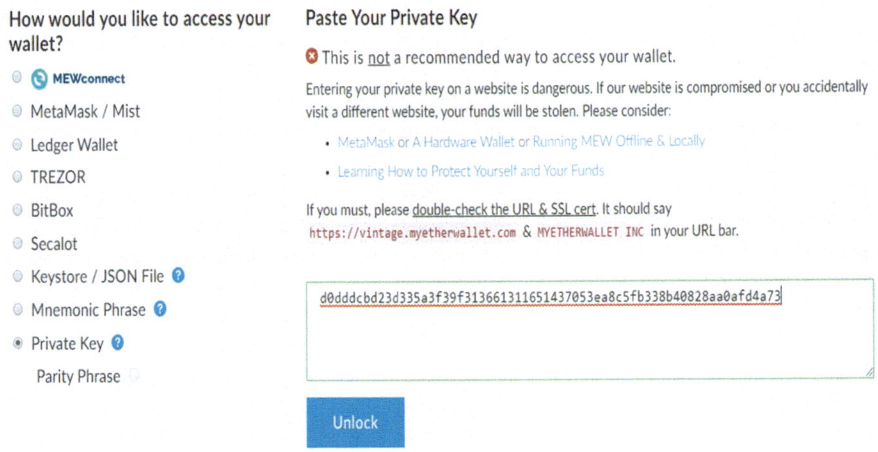

Fig. 12 Decrypt and access Wallet using a private key from one of the 10 addresses available in Ganache

The subsequent success of the first transaction gets updated in Ganache and the traces are represented in Fig. 15.

Figure 16 demonstrates the interaction with the contract pertaining to the corresponding contract address. The interacting read/write contract is represented in Fig. 17.

The list of transactions cumulated and updated on the Blockchain is depicted in Fig. 18.

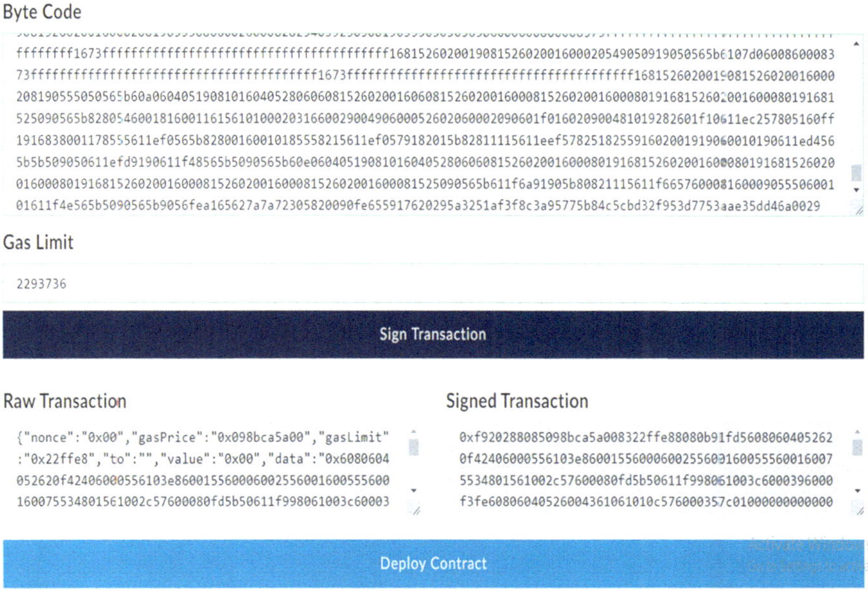

Fig. 13 Sign and deploy contract as first transaction

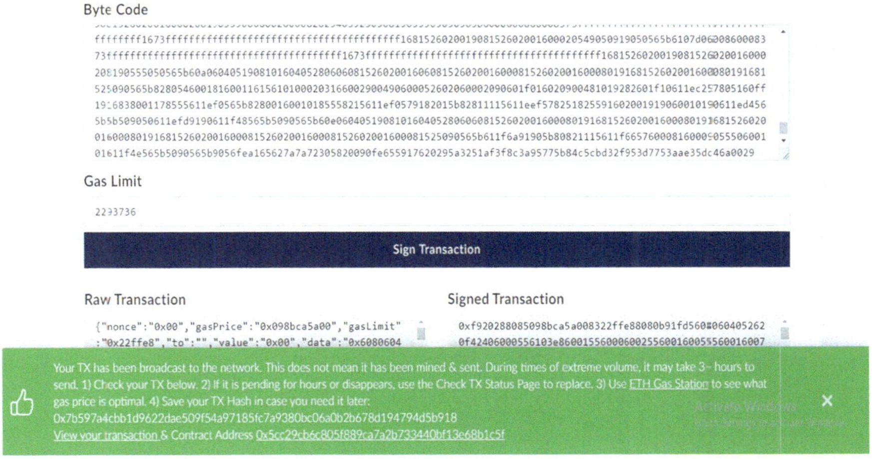

Fig. 14 Success of first transaction

Fig. 15 First transaction gets updated in Ganache

Fig. 16 Interacting with the contract

5.1 *Multi-chain Management*

Multiple or more than one contract would be required to be deployed for different contracts and relationships in order to hold up to the business confidentiality. The link between smart contracts and the stable environment of how data flows between contracts has grounds for improvement.

Standardization and interoperability

Compromising all stakeholders by adopting collaborative and international standards and information protection will be needed. Precision agriculture is a growing technique that utilizes technologies such as remote sensing, Geographic Information System, Satellite technology and Information Technology to enhance functions and services of the food and agriculture sector. This is made possible by accessing real-time data about crop growing conditions, supply chain, food safety and animal grazing. Sensor and IoT technology resolve a variety of challenges faced by Precision

Fig. 17 Interacting with the read/write contract

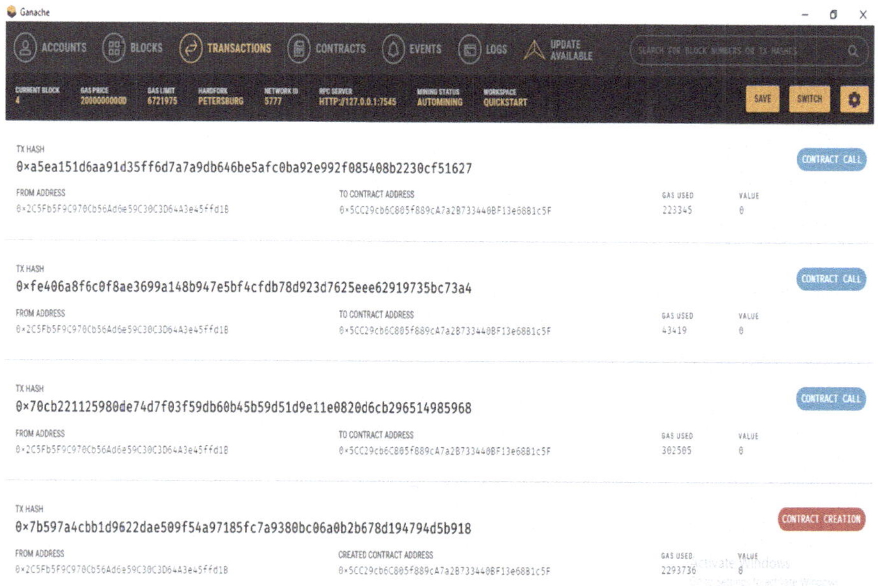

Fig. 18 The list of transactions updated on the blockchain

Agriculture. Incorporating machine learning into predicting crop growth using a self-learning model and IoT data stream can greatly increase the outcomes of Precision Agriculture.

Scalability

Blockchain's design doesn't allow large amounts of data to be stored and manipulated resulting in scalability issues which proves as a solid barrier against the integration of IoT and Blockchain.

Resolve Forking Problem

All Blockchain protocols solve this with the simple rule of considering the longest chain of blocks as the right one. This creates confusion, technical issues and fake transactions and uncertainties.

6 Conclusion

The use of Blockchain terminates tampering with data, but it cannot prevent stakeholders from adding wrong information or hiding data from the system. This shows that blockchain alone is not enough to improve the accuracy and authenticity of information in the food chain. However, when the quantity of data present is abundant, cross-checking and immutability of records might trace out frauds. This may also lower the cases of fraudulent data from entering the system.

References

1. Su S, Wang K, Kim HS (2018) Smart contract based validation for supply chain blockchain. In: IEEE international conference on Internet of Things (iThings) and IEEE Green Computing and Communications (GreenCom) and IEEE Cyber, Physical and Social Computing (CPSCom) and IEEE Smart Data (SmartData)
2. Koirala R, Dahal K, Matalonga S (2019) Supply chain using smart contract: a blockchain enabled model with traceability and ownership management. In: 9th international conference on cloud computing, data science and engineering (Confluence)
3. Salah K, Nizamuddini N, Jayaraman R, Omar M (2019) Blockchain-based soybean traceability in agricultural supply chain. IEEE Access
4. Kumar R, Tripathi R (2019) Traceability of counterfeit medicine supply chain through Blockchain. In: 11th international conference on communication systems and networks (COMSNETS)
5. Lin J, Pipattanasomporn M, Rahman S (2019) Comparative analysis of blockchain-based smart contracts for solar electricity exchanges. In: IEEE power and energy society innovative smart grid technologies conference (ISGT)
6. Nazarov AD (2019) Blockchain technology and smart contracts in the agro-industrial complex of Russia. In: IOP conference series: earth and environmental science, vol 315, pp 032016

7. Sylvester G (2018) E-Agriculture in action: Blockchain for Agriculture opportunities and Challenges. In: Food and agriculture organisation of the united nations and the international telecommunication union. Bangkok
8. Voutos Y, Drakopoulos G, Mylonas P (2014–2020) Smart agriculture: an open field for smart contracts. Regional Operational Programme, Ionian Islands
9. Ge L, Brewster C, Spek J, Smeenk A, Top J (2017) Blockchain for agriculture and food. Wageningen Economic Research and TNO
10. Torkya M, Hassaneinb AE (2020) Integrating blockchain and the internet of things in precision agriculture: analysis, opportunities, and challenges. Comput Electron Agric (Accepted)
11. Fernández-Caramés TM, Fraga-Lamas P (2018) A review on the use of blockchain for the Internet of Things. IEEE
12. Li Q-L, Yu Ma J-K, Chang Y-X, Ma F-Q, Yu H-B (2019) Markov processes in blockchain systems. Comput Soc Netw
13. Qu C, Tao M, Yuan R (2018) A hypergraph-based blockchain model and application in internet of things-enabled smart homes. Sensors
14. Shuker FM (2019) Improved blockchain network performance using hypergraph structure. J Eng Appl Sci 14:5579–5584
15. Li QL, Ma JY, Chang YX (2018) Blockchain queue theory. In: Computational dataand social networks. lecture notes in computer science bookseries (LNCS, volume 11280). Springer, pp 25–40
16. Geissler S, Prant T, Lange S, Wamer F, Hossfeld T (2019) Discrete-time analysis of the blockchain distributed ledger technology. In: 31st international teletraffic congress (ITC 31)

Chapter 5
A Deep Analysis of Gilled Mushroom Edibility Prediction Using Dimensionality Reduction Techniques

Mahendra Kumar Gourisaria, Sandali Khare, Ayush V. Patel, Manoj Sahni, and Ernesto León-Castro

1 Introduction

Mushroom or toadstool generally means fungi. It is the fruiting body member of fungi. It is consumed by people since ancient times, dating back to 3400 BCE. In ancient times, Roman Emperors used to appoint food tasters to test the mushroom for their edibility and if they are safe to eat. Thus, the cultivation of mushrooms has a long history, and commercially around twenty species of mushrooms are being cultivated [1]. There is no specific technique to determine if the mushroom is edible or poisonous. Scientifically, the mushroom is identified by its macroscopic structure. One of the ancient methods, that is used even today, works for most mushrooms; it comprises the cutting of the mushroom cap and placing the mushroom gill side down overnight, and a powdery kind of impression is formed. The color of this powdery impression helps in classifying and identifying the mushroom, and hence its edibility. Various data science experts are working on predictive models, which can efficiently predict the edibility of mushrooms. Some researchers have even used convolutional neural networks for the prediction of diseases in crops [2], nuclei segmentation [3], detecting cardiovascular diseases [4, 5], and classifying the type

M. K. Gourisaria (✉) · S. Khare · A. V. Patel
School of Computer Engineering, KIIT Deemed to Be University, Bhubaneswar, Odisha 751024, India
e-mail: mkgourisaria2010@gmail.com

M. Sahni
Department of Mathematics, School of Technology, Pandit Deendayal Energy University, Gandhinagar 382426, India
e-mail: manoj.sahni@sot.pdpu.ac.in

E. León-Castro
Faculty of Economics and Administrative Sciences, Universidad Católica de La Santísima Concepción, 4030000 Concepción, Chile

M. Sahni et al. (eds.), *Novel Developments in Futuristic AI-based Technologies*, Algorithms for Intelligent Systems, https://doi.org/10.1007/978-981-99-3076-0_5

of ECG signals. This model uses the dataset which consists of features related to mushroom caps such as the shape of the cap, cap color, and cap surface. Variables referencing the gill of mushrooms are the attachment of gill, gill spacing, gill size, and gill color. Features related to the stalk of the mushroom used are the shape of the stalk, stalk root, the surface of the stalk above the ring, the surface of the stalk below the ring, stalk color above, and stalk color below. In the end, other features used are bruises, odor, population, and habitat. The target variable used is Class which distinguishes poisoned mushrooms and edible mushrooms. In this model, we have analyzed classification techniques using evaluation metrics such as Confusion Matrix, F1 Score, and K folds Accuracy Score, on all the classification models, namely Logistic Regression, SVC, KNN, Naïve Bayes, Decision Trees, Random Forests, and ANN. We have also used the dimensionality reduction techniques such as PCA and Kernel PCA with the visualization of data after applying these for more optimization in Model I. For Model II, we have used the Pearson correlation matrix, Kendall correlation matrix, and Spearman correlation matrix to filter the features and select the highly correlated features to calculate the accuracy in the same way as Model I.

Wibowo et al. [6] implemented edible mushroom identification by classification algorithms, namely decision trees, naïve Bayes, and support vector machine, with the accuracy of 100% of SVM. In this paper, 10-cross validation was also used for refining the results [6]. Further, [7] implemented their model by using linear regression, logistic regression, naïve Bayes, random forests, SVM, and decision trees. Out of all, decision trees gave the best result in accuracy as well as reliability [7]. On the other hand, Alkronz et al. (2019) used neural networks for the implementation of their model. Their model, as shown, has come with an accuracy of 99%. The environment used by them for building a suitable network was JustNN [8]. Putra et al. [9] and Adi et al. [10] proposed a model using naïve Bayes and a voting feature interval algorithm [9, 10].

2 Materials and Methods

This section comprises the study of the dataset and pre-processing of the dataset and our analysis overview. An analysis is done by using the techniques such as dimensionality reduction and grid search CV. An overview of the same is shown in Fig. 1.

2.1 Dataset

The dataset is taken from UCI in CSV format. Based on morphological characters, the type of mushroom is classified. The dataset contains 8124 rows and 23 columns

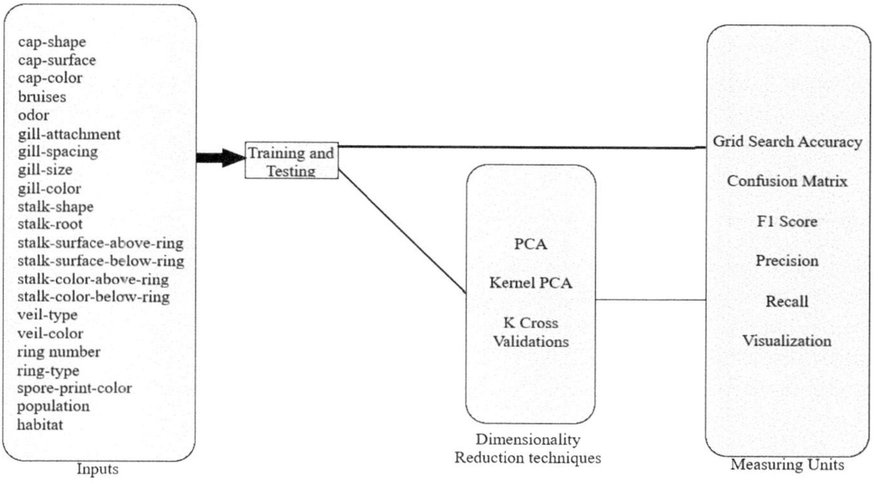

Fig. 1 Approach overview

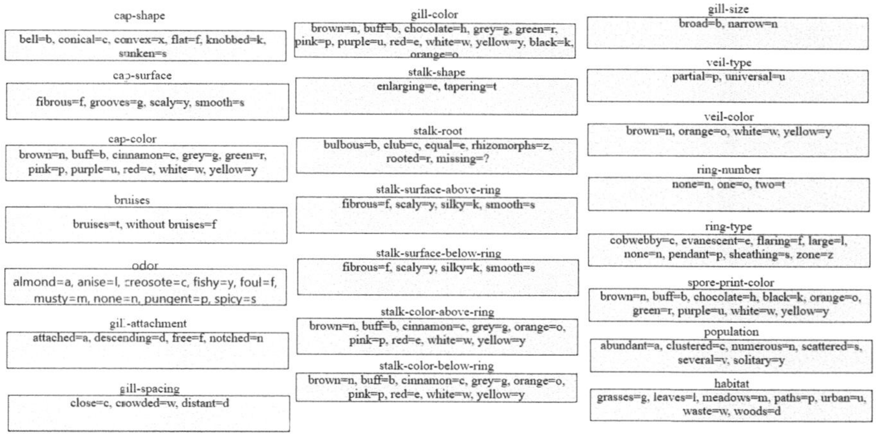

Fig. 2 Overview of the dataset

out of which 21 columns are of string data type and 2 columns are of a Boolean data type; description of the same with their categorical values is given in Fig. 2.

2.2 *Pre-processing the Data*

The software used is Jupyter Notebook (Anaconda) for this model. The dataset we are using is clean, so there are no outliers or missing values. In this paper, categorical values are converted to ordinal values using Label Encoder. It converts labels to a numeric form to make them readable by a machine. Further, Standard Scaler is used to remove the mean and scale each variable to unit variance. To pre-process the data, we are using different types of correlation matrices to detect the highly correlated features and select them for Model II. Different types of correlation techniques are described below.

Spearman Correlation

Spearman Correlation is known for determining the relationship between two ordinal or continuous variables. Variables are monotonous, i.e., they are likely to change together but not necessarily at the same rate. This correlation is specifically based on ordinal data rather than raw data [11]. Its coefficient is called a coefficient of Pearson correlation between ranked variables. It can be defined in Eq. 1 as follows:

$$r_s = \rho_{rgX,rgY} = \frac{\text{cov}(rg_X, rg_Y)}{\sigma_{rg_X}\sigma_{rg_Y}}. \tag{1}$$

Here, let X_i, Y_i be the raw scores of the data of n samples which are then converted to ranks rgX_i, rgY_i. In this equation, ρ denotes the coefficient of Pearson correlation, $\text{cov}(rg_X, rg_Y)$ stands for the covariance of ranked data, and $\sigma_{rg_X}\sigma_{rg_Y}$ denotes the standard deviation of the ranked data [12].

Pearson Correlation

This is the most common correlation technique which is used to determine the linear relationship between two data. Its coefficient is called the Pearson Product Moment Correlation Coefficient (PPMCC) [13]. It can be formulated in Eq. 2 as

$$\rho_{X,Y} = \frac{\text{cov}(X, Y)}{\sigma_X, \sigma_Y}. \tag{2}$$

Here, X and Y represent a pair of random variables, cov represents the calculation of covariance of X and Y, and $\sigma_X\sigma_Y$ represents the standard deviation of X and Y [10].

Kendall Correlation

Kendall Correlation is the alternative to Spearman Correlation as both of them determine the relationship between two ranked variables. The coefficient of the Kendall Rank Coefficient is Kendall's Tau coefficient. It is generally used with a small sample or tied ranks [14]. It can be formulated as in Eq. 3 as follows:

$$\tau = \frac{C - D}{C + D}. \tag{3}$$

Here, C determines the concordant pairs and D stands for the discordant pairs.

Another technique used for the pre-processing of data is Dimensionality reduction. Dimension Reduction refers to the reduction of the dimensions of the features. In this paper, two types of dimensional reduction techniques are used, which are mentioned below.

Principal Component Analysis (PCA)

PCA is a technique of dimensionality reduction. It belongs to the group of factor analysis. It is mainly used to minimize the dimensions of a dataset that contains highly correlated or lightly correlated features [15, 16]. It retains the variation present up to the maximum extent [17]. For the analysis, two-dimensional samples are constructed with axes as factors. N component parameter is used which tells the number of new variables which are constructed as a mixture of the initial variables. In our model, we have used n components as 2. To calculate the PCA, the steps to be followed are mentioned below.

Step 1: We calculate the mean of each column as per Eq. 4:

$$\text{mean} = \sum \frac{x_i}{n}, \tag{4}$$

where x_i is the sum of the column and n gives the total number of values.

Step 2: Center the columns by subtracting column means.

Step 3: Covariance is calculated for the centered matrix.

Step 4: Perform Eigendecomposition of the covariance matrix which is formulated as in Eq. 5:

$$\lambda_i = \frac{1}{n} \sum_{N=1}^{n} \left(e_i^T \omega_N\right)^2, \tag{5}$$

where ω_i represents the mean-centered data.

Kernel Principal Component Axis (Kernel PCA)

This is the extension of PCA. The techniques used are kernel methods. Linear techniques that were used originally in PCA are supervised by a kernel, and they are performed by a kernel Hilbert space which is reproducing in nature [18]. In our model, the kernel used is RBF, and the n components considered are 2. It can be formulated as in Eq. 6:

$$\lambda v = Cv \Leftrightarrow \lambda(x_k \cdot v) = (x_k \cdot C_v), \forall k = 1, 2, \ldots, N. \tag{6}$$

Here, C stands for the covariance matrix. This equation can be represented in another form of dot product space F, which is the other form of Eq. 7 as

$$\phi : R^d \rightarrow F, x \mapsto \dot{x}. \tag{7}$$

K-fold Cross-Validation

Cross-validation refers to the resampling process and evaluates the model on a smaller data sample than the original one. The process involves dividing the data into a random k number of folds that are roughly the same size. One of the folds is used for testing, and the rest folds are used to train the data, which are k-1 folds. The whole procedure keeps on repeating k times, and every time a different fold is used for the validation of the data [19]. The K fold cross-validation error is given by Eq. 8 as

$$CV_{(k)} = \frac{1}{k}\sum_{i=1}^{k} MSE_i, \tag{8}$$

where MSE stands for mean squared error.

3 Classification Methods

This section of the paper comprises a brief discussion of different classification models used for the analysis.

3.1 Logistic Regression (LR)

Logistic Regression or Logit Regression is a technique of classification that is used with binary data also called categorical data [20]. The different types of LR are Binary Logistic Regression, Multinomial Logistic Regression, and Ordinal Logistic Regression. The value of this regression must bind under 0–1. In the research paper, for hyperparameter tuning, the parameters used are the C index and penalty. C stands for Concordance; it tells the area under a ROC curve. It is a measure of the goodness of fit of the binary results of a logistic regression model. The penalty is used for imposing a penalty on a logistic regression model based on too many variables. Table 1 shows in detail the values used for the same. The general formula of Logistic Regression is based on a sigmoid function [21]. It can be formulated as in Eq. 9:

$$P = \frac{e^{a+bx}}{1 + e^{a+bx}}. \tag{9}$$

Here, P is the target variable and its value varies from 0 to 1.

Table 1 Hyperparameter name versus Hyperparameter values w.r.t. Logistic Regression

Hyperparameter name	Hyperparameter value
C	1.0
Penalty	12
Solver	False

3.2 *Decision Trees (DT)*

The decision tree algorithm is one of the supervised algorithms, therefore it requires a label along with functioning. In the Decision Tree method, data is split regularly under certain parameters [22]. This model is used for both Regression data and Categorical data [23]. In this paper, it is used with categorical data as our target variable is categorical, and therefore it predicts the percentage or probability [24]. Parameters tuned for this model in the research paper are random state and max depth. Max depth is used as a limit for stopping the further split of nodes during building the initial tree. Table 2 shows in detail the values used for the same. Some important terms for Decision Trees are Leaf Node, Root Node, Decision Node, Parent Node, Child Node, Splitting, and Sub-Tree/Branch [25]. Figure 3 shows the overview of these terms.

Table 2 Hyperparameter name versus Hyperparameter values w.r.t. Decision Tree

Hyperparameter name	Hyperparameter value
Random_state	10
Max_depth	6
Criterion	Gini

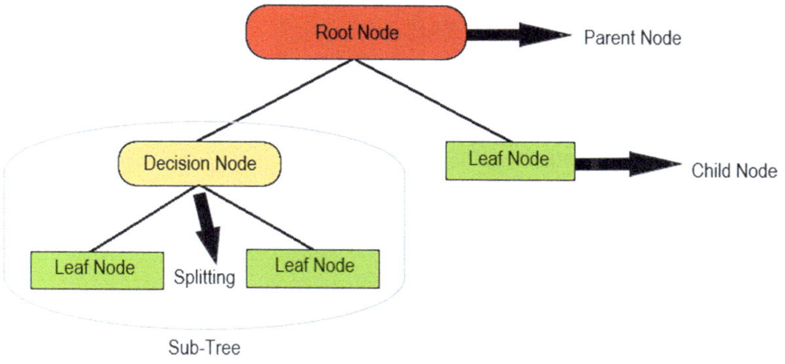

Fig. 3 Overview of the decision tree

3.3 Naïve Bayes (NB)

Naïve Bayes is the most popular technique used with classification models [26]. It comes under supervised learning. It is said to be a complicated technique. Its main concept is to make predictions to classify by using probability [27]. NB algorithms are the collection of approved Bayes theorem-based algorithms [28] and are considered a generative model [21]. A naïve Bayes classifier is completely related to the Bayes theorem. Bayes theorem can be mathematically stated as in Eq. 9:

$$P(Y/X) = \frac{P(X/Y)P(Y)}{P(X)}. \tag{10}$$

Here, P (Y|X) states the probability of Y considering event X is true, P(Y) is the priority of Y, and P (X|Y) is stated as the posterior probability of X [29]. NB can be formulated as

$$\hat{Y} = \arg\max_{Y} P(Y) \prod_{i=1}^{N} P(X_i/Y). \tag{11}$$

3.4 Random Forest (RF)

It is a supervised technique. The base of random forests is decision trees [30]. This technique is used for the classification model as well as the regression model. It works with a categorical target variable for the classification model. Multiple decision trees are combined to form a random forest. Several samples are taken randomly comprising different decision trees. Each decision tree gives its prediction, and the best prediction is selected by the random forest algorithm [31]. In this paper, a random forest classifier is used, which divides the training set into randomly selected subsets and performs a decision tree algorithm on each subset, then it aggregates the voting of the decision trees and with this, the final class of the test object is decided [32]. It is clearly shown in Fig. 4. Tuned parameters are n_estimators, max depth, and random state. Max depth is used for representing the depth of each tree. The more deep the tree is, the more information is captured about the data. But parallelly, it created the case of overfitting with our dataset. N estimators speak of the total number of trees in the forest. Table 3 shows in detail the values used for the same.

Fig. 4 Overview of RF classifier

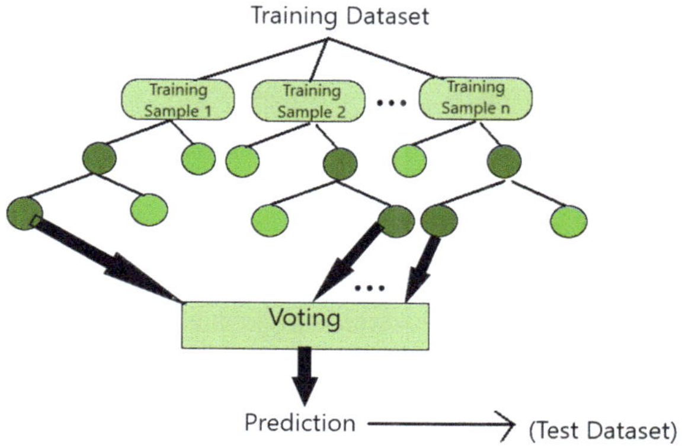

Table 3 Hyperparameter name versus Hyperparameter values w.r.t. Random Forest

Hyperparameter name	Hyperparameter value
N_estimators	100
Max_depth	4
Random_state	2

3.5 Support Vector Machine (SVM)

SVM is a useful linear model [33, 34]. It is a supervised technique. It is used for accomplishing both linear and nonlinear problems. It works on a very smart yet critical idea [35]. The kernel trick technique is used in transforming the data; it finds the optimal boundary between feasible outputs based on these transformations. It creates a hyperplane or a line and separates the mixed data into classes. An overview of SVM is shown in Fig. 5. The parameters tuned are kernel and random state. Radial Basis Kernel is used for finding regression lines or nonlinear classifiers. Table 4 shows in detail the values used for the same.

Fig. 5 Overview of SVM

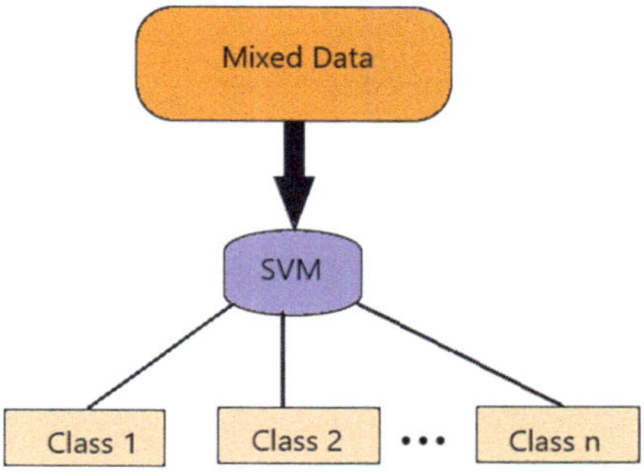

Table 4 Hyperparameter name versus Hyperparameter values w.r.t. SVM

Hyperparameter name	Hyperparameter value
Kernel	RBF
Random_state	42

3.6 K-Nearest Neighbor (KNN)

KNN algorithm is a useful, simple, and supervised machine learning technique. It is used to solve both regression and classification problems. It is also called lazy learning, a non-parametric technique. This algorithm works by finding the K-nearest matches in the training set and then using the closest batch level for prediction [36]. In the paper, default parameters are used for KNN as they were giving the best results.

3.7 Artificial Neural Network (ANN)

ANN refers to computational networks. This algorithm is based on the human neuron model. Neural networks have decision-making abilities just like biological neurons [37]. The model learns with the help of data while it is being designed. Modeling of connections is done as weights [38]. A negative connection is called an inhibitory connection, whereas a positive connection is referred to as an excitant connection. There are 3 layers in a simple neural network system; the layers are hidden layer, the input layer, and the result layer [4]. We apply the activation function in the hidden layer. In this model, we have used a sigmoid and ReLU activation functions. The input layer consists of inputs given as weights which are then summed up. In the first hidden layer and second hidden layer, activation is ReLU and units are 46; the kernel initializer for both is uniform. In the output layer, activation is sigmoid and the unit is 1. The loss function we used is binary cross-entropy, and we have used the Adam optimizer while compiling the ANN model. Table 5 shows in detail the values used for the same.

Table 5 Hyperparameter name versus Hyperparameter values w.r.t. ANN

Hyperparameter name	Hyperparameter value
loss	Binary_crossentropy
Optimizer	adam
Input_dim	95

4 Evaluation Metrics

Grid Search is another name for exhaustive search. Grid Search Accuracy (GSA) goes through all the unions of hyperparameters, and all the defined hyperparameter values will be given a try. The best accuracy from all the accuracies will be selected [39]. CV in grid search refers to several cross-validations that are tired of all the selected sets of hyperparameters. In this model, we have used 5 as the CV value.

Confusion Matrix is used specifically for classification problems. This table is used for describing the performance of the testing data when the true value is known. It is an N x N matrix where N determines the number of target classes. Figure 6 gives an overview of the confusion matrix [40]. Here, if the true class and predicted class match with the positive value, then it is called TP. If the predicted class predicts the value as positive but in the true class it is negative, then it is referred to as FP. If the value in the actual class is positive but is predicted negative, then it is called FN. Lastly, if the value in the true class is negative and the predicted class also predicts negative, then it is called TN.

Precision or positive predictive value refers to the ratio of true positive predicted observations by total positive predicted observations. In other words, it measures the relevancy of the result and is given in Eq. 12 [41]:

$$\text{Precision} = \frac{TP}{(TP + FP)}. \tag{12}$$

Recall also called sensitivity refers to the ratio of total observations that were positive or retrieved. In other words, it measures the number of truly relevant results that are returned. It is also called the TP rate and is given in Eq. 13 [42]:

$$\text{Recall} = \frac{TP}{(TP + FN)}. \tag{13}$$

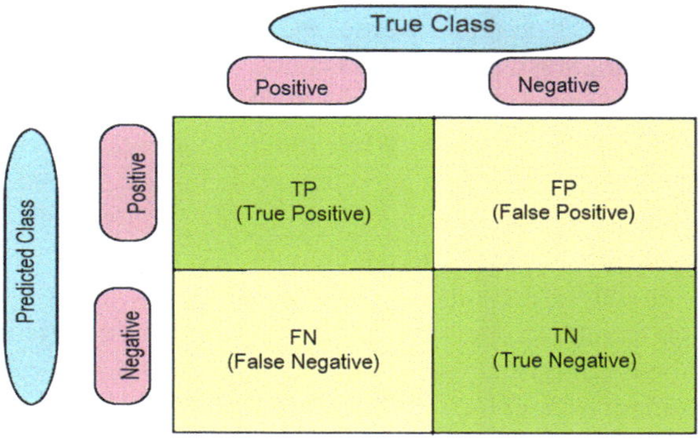

Fig. 6 Overview of the confusion matrix

F1 Score refers to the weighted harmonic mean of recall and precision of the test and is given in Eq. 14. In other words, it refers to the weighted average of recall and precision [43]:

$$F1\,\text{Score} = 2 \times \frac{\text{Precision} \times \text{recall}}{\text{Precision} \times \text{recall}} = \frac{TP}{TP + \frac{1}{2}(FP + FN)'}. \tag{14}$$

Visualization of PCA and kernel PCA reduction techniques can be obtained in machine learning models.

5 Results and Discussion

In our model, various evaluation metrics have been used over different classification techniques for all the models. The result of them is mentioned in the table below.

Figure 7 shows the matrix obtained by the Spearman correlation. Here, it is seen that the features which are highly correlated with the target variable are bruises, gill size, and gill color. By highly correlated, it means that features are related to the target variable in either positive correlation or negative correlation. Bruises, gill size, and gill color are all positively correlated with correlation coefficients of 0.5, 0.54, and 0.53, respectively. We also checked the correlation between these three features, and they were not correlated, which is why none of them is dropped.

Figure 8 shows the Pearson correlation matrix obtained. As observed in the figure, it is giving similar results to the Spearman correlation.

Figure 9 shows the matrix obtained by the Kendall Correlation technique. Here, it can be observed that bruises and gill size are the only features that are highly correlated with the target variable, with a correlation coefficient of 0.5 and 0.54, respectively.

In Table 6, the results for the model which consists of all the features are shown. The results are shown for three modeling approaches, namely PCA and cross-validation technique, Kernel PCA and cross-validation, and the last one without any reduction techniques. It is observed that with no reduction technique of features, we are getting the best results. KNN and ANN models are giving the best results in both test and train sets. While other models gave average results, Linear Regression gave good results with no reduction technique applied. It gives similar results in both PCA and Kernel PCA reduction models. Similarly, all the other models give the same results for both reduction techniques.

In Table 7, results are displayed for the features selected by the Kendall correlation technique. Features selected by Kendall correlation are only used. Results are shown concerning the two modeling approaches, reduction with PCA and cross-validation technique and another one with Kernel PCA reduction technique with cross-validation. In this model, we are getting average results from all the models ranging from 75 to 79% for both test and train sets.

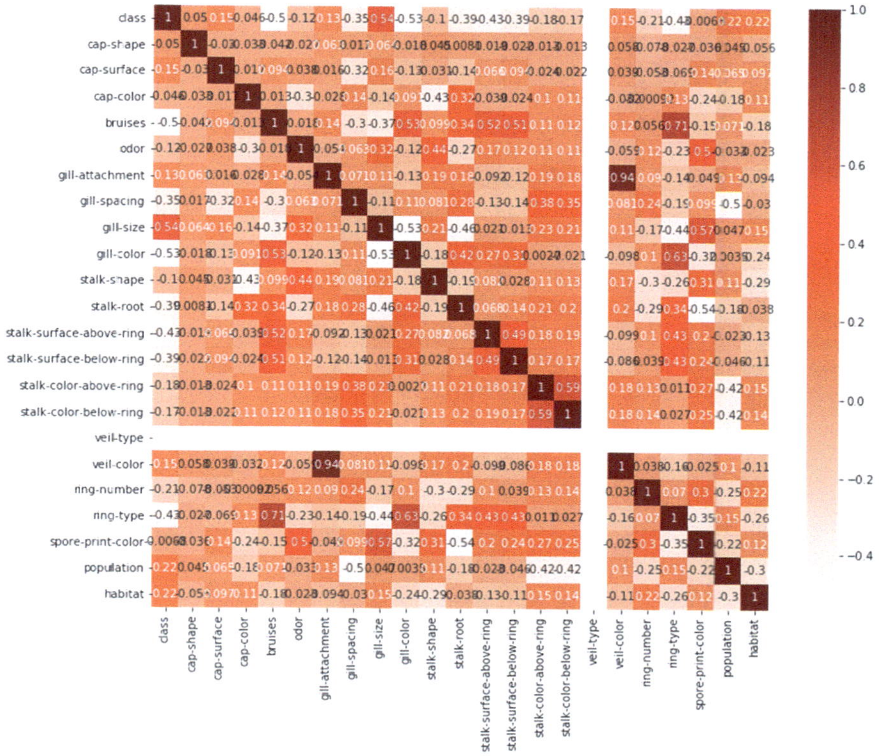

Fig. 7 Spearman correlation

In Table 8, results are displayed for selected features by Spearman and Pearson correlation techniques. As both of these techniques were giving the same correlated features, the result of them is calculated together. Results are shown concerning the two modeling approaches, reduction with PCA and cross-validation technique and another one with Kernel PCA reduction technique with cross-validation. This model is giving the worst results with an accuracy of around 30% for all the models for the test set, and for the train set, it is the case of overfitting with the accuracy ranging from 75 to 90%.

In this paper, the visualization of all the models is done concerning both dimensionality reduction techniques, PCA and Kernel PCA. In Figs. 10 and 11, visualization of PCA and kernel PCA respectively for concerning models with all the features is shown.

In Figs. 12 and 13, visualization of PCA and kernel PCA respectively for concerning models with Kendall correlated features is shown.

In Figs. 14 and 15, visualization of PCA and kernel PCA respectively with Spearman-Pearson correlation is shown.

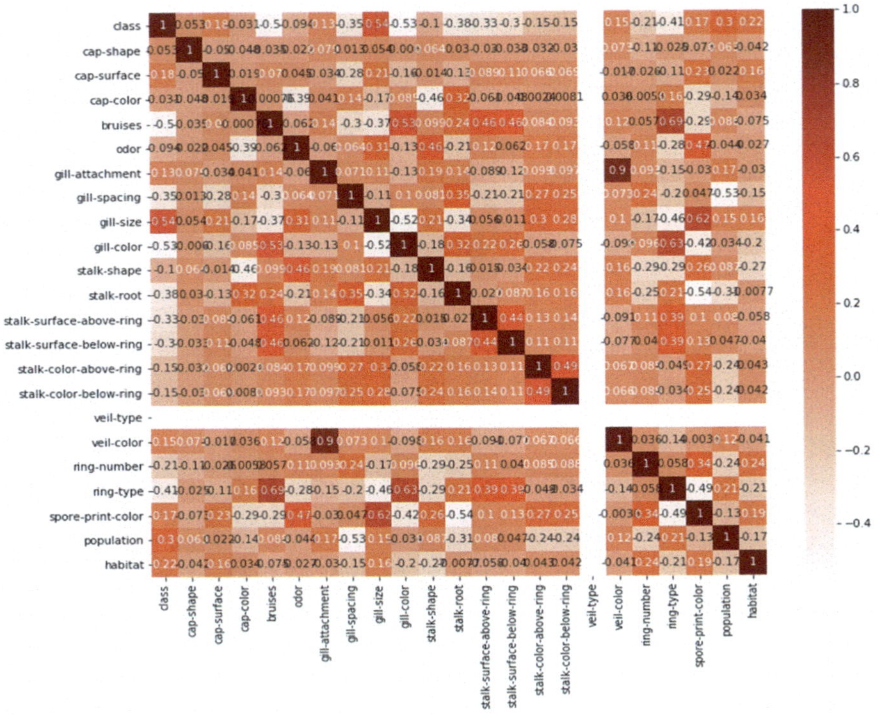

Fig. 8 Pearson correlation

6 Conclusion and Future Work

Cultivating Mushrooms is a widespread business. It is increasing rapidly. As of now, there are no specific techniques to identify edible mushrooms from the set of mushrooms. This paper draws attention to the prediction of the edibility of mushrooms through different classification techniques. In this paper, we used a dataset from Kaggle and did a classification analysis of the dataset. Techniques we used are logistic regression, naïve Bayes, KNN, ANN, decision trees, random forest, and SVM. For further analysis, we have also used the K cross-validation along with the reduction dimensionality methods that are PCA and kernel PCA for refining the results. For the evaluation, we used grid search accuracy for more precise results. All the techniques gave decent results, but ANN and KNN in Model I gave good results with an accuracy of almost 100%. It can be concluded that for analyzing the edibility of mushrooms, both macromorphological and micromorphological features should be considered.

For further research, a mobile application can be developed by the researchers to identify the edible and wild mushrooms visually [44], by using the Convolutional

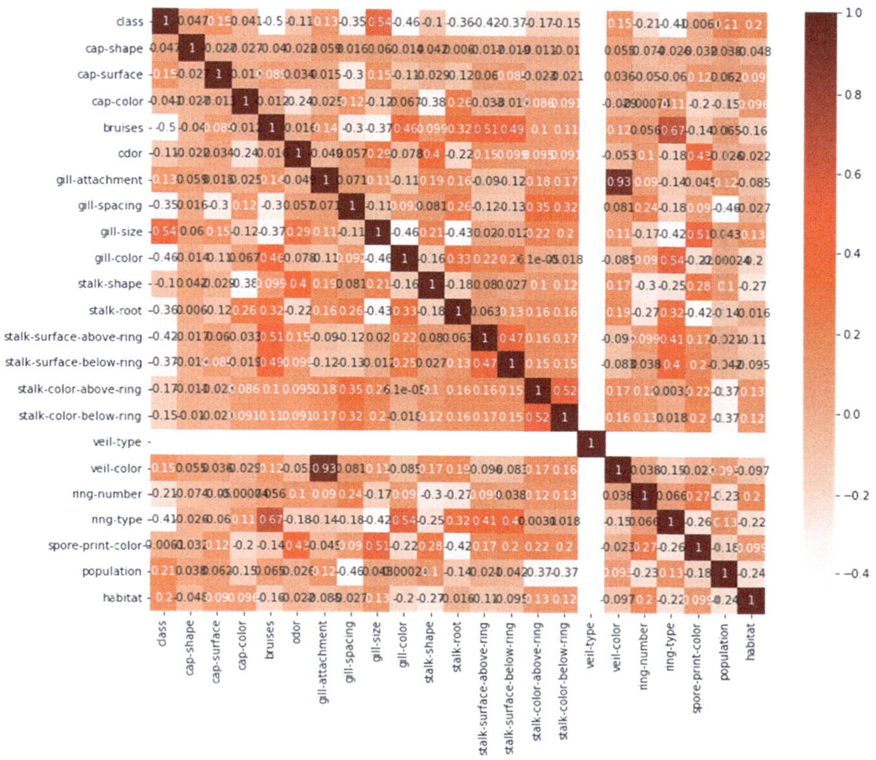

Fig. 9 Kendall correlation

Neural Networks (CNN) technique applied to the dataset [45], or by using cloud computing, the photo of the mushroom can be shown in the mobile [46].

Table 6 Comparison of a result obtained by different techniques for all features

Techniques	Models	GSA	Precision	Recall	F1 in test	F1 in train	Confusion matrix	
PCA and K-fold cross-validation	LR	0.90	0.92	0.91	0.91	0.90	803	17
							134	167
	DT	0.92	0.93	0.92	0.92	0.93	801	16
							113	692
	NB	0.89	0.91	0.90	0.89	0.90	801	19
							140	665
	RF	0.91	0.92	0.91	0.91	0.91	802	18
							126	679
	KNN	0.93	0.93	0.93	0.93	0.94	794	26
							95	710
	SVM	0.91	0.92	0.91	0.91	0.91	809	11
							129	676
	ANN	0.92	0.93	0.92	0.92	0.93	795	25
							98	707
Kernel PCA and K-fold cross-validation	LR	0.89	0.91	0.89	0.89	0.89	818	2
							171	634
	DT	0.89	0.92	0.92	0.92	0.92	749	71
							55	750
	NB	0.89	0.91	0.88	0.88	0.89	818	2
							184	621
	RF	0.89	0.91	0.89	0.89	0.89	820	0
							181	624
	KNN	0.95	0.94	0.94	0.94	0.96	774	46
							55	750
	SVM	0.91	0.92	0.91	0.90	0.91	807	13
							127	678
	ANN	0.93	0.93	0.93	0.93	0.94	794	26
							89	716

Table 7 Results for the model consisting of correlated features through Kendall correlation

Techniques	Models	GSA	Prec	Rec	F1 in test	F1 in train	Confusion matrix	
PCA and K-fold cross-validation	LR	0.76	0.79	0.75	0.74	0.75	1162	95
							509	672
	DT	0.76	0.78	0.76	0.75	0.76	776	481
							109	1072
	NB	0.77	0.78	0.76	0.75	0.75	776	481
							109	1072
	RF	0.76	0.79	0.76	0.75	0.76	776	481
							109	1072
	KNN	0.75	0.79	0.75	0.74	0.75	1162	95
							509	672
	SVM	0.77	0.78	0.76	0.75	0.76	776	481
							109	1072
	ANN	0.76	0.78	0.76	0.75	0.79	776	481
							109	1072
Kernel PCA and K-fold cross-validation	LR	0.76	0.78	0.76	0.75	0.76	776	481
							109	1072
	DT	0.77	0.78	0.76	0.75	0.76	776	481
							109	1072
	NB	0.77	0.78	0.76	0.75	0.76	776	481
							109	1072
	RF	0.77	0.78	0.76	0.75	0.75	776	481
							109	1072
	KNN	0.75	0.79	0.75	0.74	0.75	1162	95
							509	672
	SVM	0.77	0.78	0.76	0.75	0.76	776	481
							109	1072
	ANN	0.76	0.78	0.76	0.75	0.79	776	481
							109	1072

Table 8 Result for a model consisting of selected features by Spearman and Pearson correlation technique

Techniques	Models	GSA	Prec	Rec	F1 in test	F1 in train	Confusion matrix	
PCA and K-fold cross-validation	LR	0.20	0.20	0.20	0.20	0.77	341	916
							1029	152
	DT	0.31	0.31	0.31	0.31	0.69	411	846
							831	350
	NB	0.22	0.22	0.22	0.22	0.79	327	930
							964	217
	RF	0.31	0.31	0.31	0.31	0.85	465	792
							890	291
	KNN	0.33	0.33	0.33	0.33	0.86	521	736
							890	291
	SVM	0.34	0.33	0.34	0.33	0.85	537	720
							890	291
	ANN	0.35	0.34	0.35	0.34	0.90	546	711
							876	305
Kernel PCA and K-fold cross-validation	LR	0.77	0.76	0.75	0.75	0.76	1095	162
							446	735
	DT	0.86	0.80	0.79	0.79	0.86	879	378
							133	1048
	NB	0.74	0.79	0.73	0.72	0.73	1195	62
							586	595
	RF	0.87	0.84	0.84	0.84	0.86	996	261
							133	104
	KNN	0.88	0.81	0.81	0.81	0.87	1070	187
							271	910
	SVM	0.83	0.84	0.84	0.84	0.84	1020	237
							165	1016
	ANN	0.84	0.84	0.84	0.84	0.87	1006	251
							149	1032

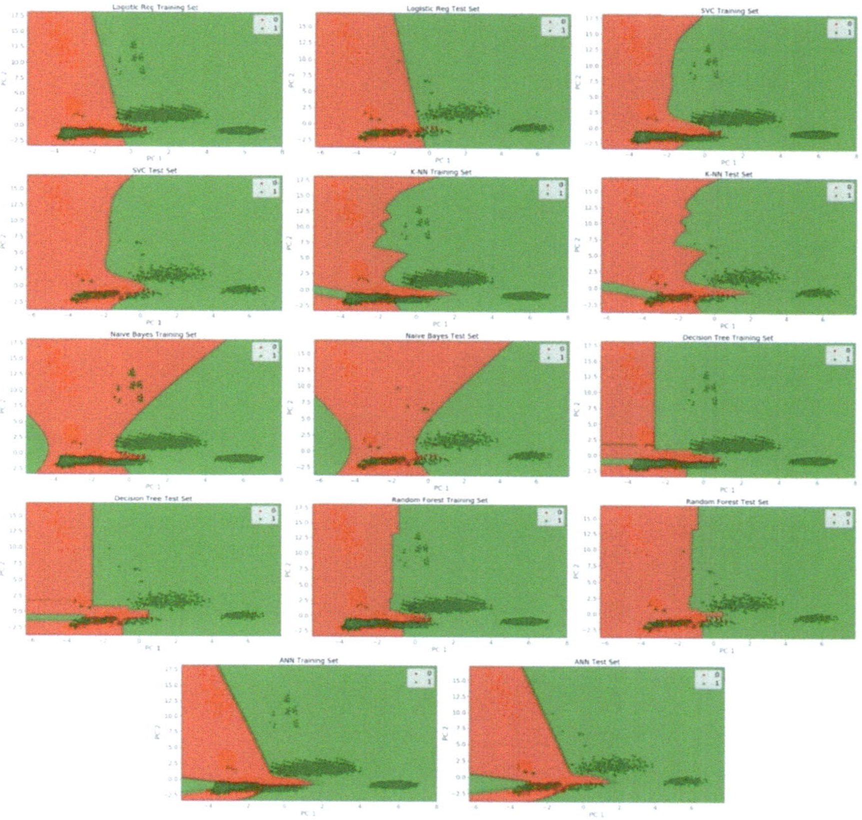

Fig. 10 Visualization for all features w.r.t. PCA

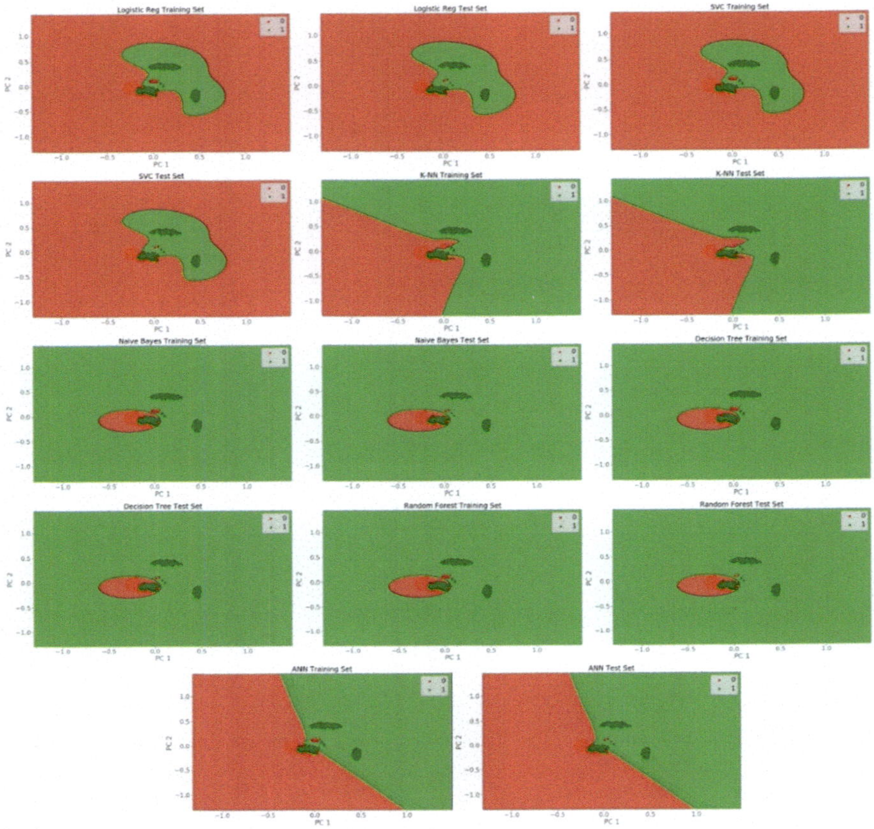

Fig. 11 Visualization for all features w.r.t. kernel PCA

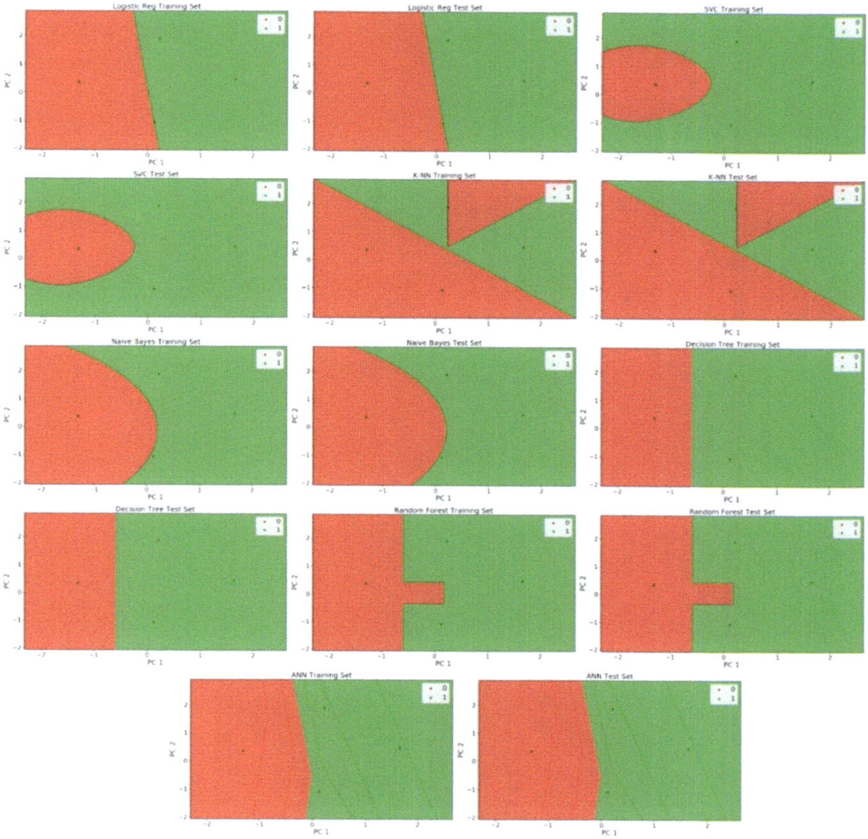

Fig. 12 Visualization of PCA w.r.t. Kendall correlation

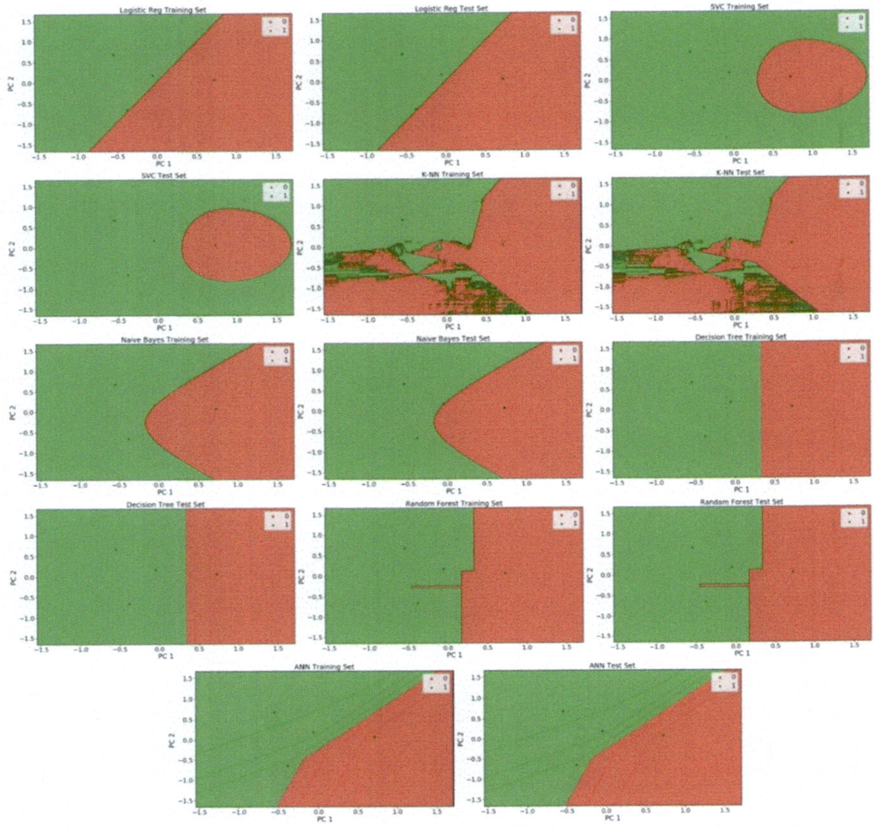

Fig. 13 Visualization of kernel PCA w.r.t. Kendall correlation

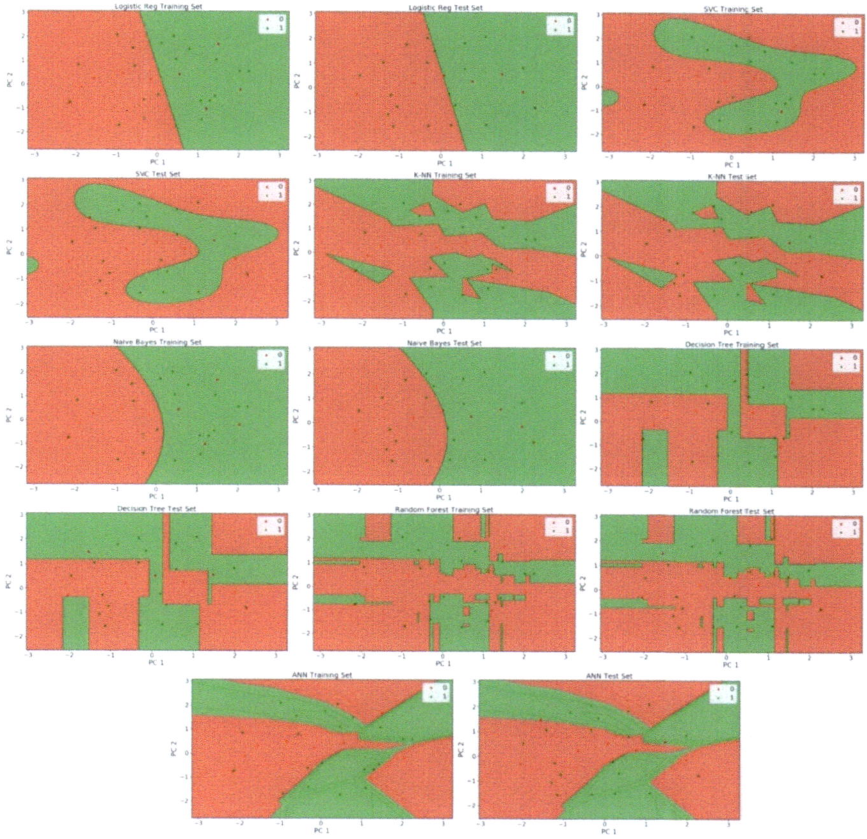

Fig. 14 Visualization of PCA w.r.t. Spearman and Pearson correlation

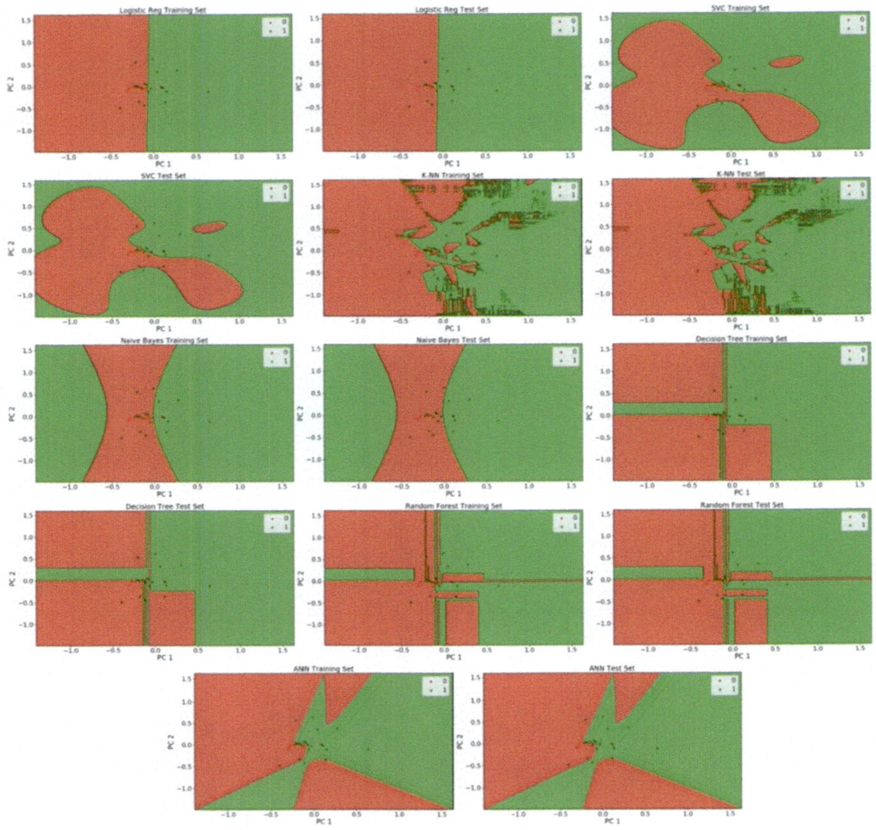

Fig. 15 Visualization of kernel PCA w.r.t. Spearman and Pearson correlation

References

1. Wikipedia. https://en.wikipedia.org/wiki/Edible_mushroom. Last accessed 1 Oct 2020
2. Das S, Sharma R, Gourisaria MK, Rautaray SS, Pandey M (2020) A model for probabilistic prediction of paddy crop disease using convolutional neural network. In: Intelligent and cloud computing. Springer, Singapore, pp 125–134
3. Dey S, Gourisaria MK, Rautray SS, Pandey M (2021) Segmentation of nuclei in microscopy images across varied experimental systems. In: Satapathy S, Zhang YD, Bhateja V, Majhi R (eds) Intelligent data engineering and analytics. Advances in intelligent systems and computing, vol 1177. Springer, Singapore
4. Anand A, Anand H, Rautaray SS, Pandey M, Gourisaria MK (2020) Analysis and prediction of chronic heart diseases using machine learning classification models. Int J Adv Trends Comput Sci Eng 9(5):8479–8487
5. Soni J, Ansari U, Sharma D, Soni S (2011) Predictive data mining for medical diagnosis: an overview of heart disease prediction. Int J Comput Appl 17(8):43–48
6. Wibowo A, Rahayu Y, Riyanto A, Hidayatulloh T (2018) Classification algorithm for edible mushroom identification. Int Conf Inf Commun Technol (ICOIACT) 2018:250–253

7. Chelliah BJ, Kalaiarasi S, Anand A, Janakiram G, Rathi B, Warrier NK (2018) Classification of mushrooms using supervised learning models. Int J Emerg Technol Eng Res (IJETER) 6(4)
8. Alkronz ES, Moghayer KA, Meimeh M, Gazzaz M, Abu-Nasser BS, Abu-Naser SS (2019) Prediction of whether mushroom is edible or poisonous using back-propagation neural network
9. Putra BM (2008) Klasifikasi Jamur ke Dalam Kelas Dapat Dikonsumsi Atau Beracun Menggunakan Algoritma VFI 5 (Studi Kasus: Famili Agaricus dan Lepiota). IPB, Bogor, Laporan Akhir
10. Adi G, Pradana S (2016) Identifikasi jamur beracun pada jenis jamur famili agaricus dan lepiota berdasarkan klasifikasi. Univeritas Nusantara PGRI Kediri, Kediri, Laporan Akhir
11. Minitab. https://support.minitab.com/en-us/minitab-express/1/help-and-how-to/modeling-statistics/regression/supporting-topics/basics/a-comparison-of-the-pearson-and-spearman-correlation-methods/date3October.
12. Wikipedia. https://en.wikipedia.org/wiki/Spearman%27s_rank_correlation_coefficient. Last accessed 4 Oct 2020
13. Staticshowto. https://www.statisticshowto.com/probability-and-statistics/correlation-coefficient-formula/. Last accessed 3 Oct 2020
14. Towardsdatascience. https://towardsdatascience.com/kendall-rank-correlation-explained-dee01d99c535. Last accessed 4 Oct 2020
15. Granato D, Santos JS, Escher GB, Ferreira BL, Maggio RM (2018) Use of principal component analysis (PCA) and hierarchical cluster analysis (HCA) for multivariate association between bioactive compounds and functional properties in foods: a critical perspective. Trends Food Sci Technol 72:83–90
16. Ezire. https://www.dezyre.com/data-science-in-pythontutorial/principal-component-analysis-tutorial. Last accessed 4 Oct 2020
17. Sharma R, Das S, Gourisaria MK, Rautaray SS, Pandey M (2021) A model for probabilistic prediction of paddy crop disease using convolutional neural network. In: Intelligent and cloud computing. Springer, Singapore, pp 125–134
18. Schölkopf B, Smola A, Müller KR (1997) Kernel principal component analysis. In: International conference on artificial neural networks. Springer, Berlin, pp 583–588
19. Rodriguez JD, Perez A, Lozano JA (2009) Sensitivity analysis of k-fold cross validation in prediction error estimation. IEEE Trans Pattern Anal Mach Intell 32(3):569–575
20. Khemphila A, Boonjing V (2018) Comparing performances of logistic regression, decision trees, and neural networks for classifying heart disease patients. In: International conference on computer information systems and industrial management applications (CISIM), pp 193–198
21. Harshvardhan GM, Gourisaria MK, Pandey M, Rautaray SS (2020) A comprehensive survey and analysis of generative models in machine learning. Comput Sci Rev 38:100285
22. Krishnaiah V, Narsimha G, Chandra NS (2013) Diagnosis of lung cancer prediction system using data mining classification techniques. Int J Comput Sci Inf Technol 4:39–45
23. Singh V, Gourisaria MK, Das H (2021) Performance analysis of machine learning algorithms for prediction of liver disease. In: 2021 IEEE 4th international conference on computing, power and communication technologies (GUCON), pp 1–7
24. Tso GK, Yau KK (2007) Predicting electricity energy consumption: a comparison of regression analysis, decision tree, and neural networks. Energy 32(9):1761–1768
25. Brijain M, Patel R, Kushik M, Rana K (2014) A survey on decision tree algorithm for classification. Int J Eng Dev Res 2(1)
26. Singh V, Agrawal R, Gourisaria MK, Singh PK, Das H (2022) Comparative analysis of machine learning models for early detection of fetal disease using feature extraction. In: 2022 IEEE 11th international conference on communication systems and network technologies (CSNT), pp 169–175
27. Vijararani S, Dhayanand S (2015) Data mining classification algorithms for kidney disease prediction. Int J Cybern Inform 4(4):13–25
28. Subbalakshmi G, Ramesh K, Rao C (2011) Decision support in heart disease prediction system using Naïve Bayes. Indian J Comput Sci Eng 2(2):170–176

29. Saritas MM, Yasar A (2019) Performance analysis of ANN and Naïve Bayes classification algorithm for data classification. Int J Intell Syst Appl Eng 7(2):88–91
30. Casanova R, Saldana S, Chew EY, Danis RP, Greven CM, Ambrosius WT (2014) Application of random forests methods to diabetic retinopathy classification analyses. PLOS One
31. Masetic Z, Subasi A (2016) Congestive heart failure detection using random forest classifier. Comput Methods Programs Biomed 130:54–64
32. Rodriguez-Galiano V, Sanchez-Castillo M, Chica-Olmo M, Chica-Rivas M (2016) Machine learning predictive models for mineral prospectivity. An evaluation of neural networks, random forest, regression trees, and support vector machines 71:804–818
33. Mythili T, Mukerji D, Padalia N, Naidu A (2013) A heart disease prediction model using SVM decision trees-logistic regression (SDL). Int J Comput Appl 68:11–15
34. Yu W, Liu T, Valdez R, Gwinn M, Khoury MJ (2010) Application of support vector machine modeling for prediction of common diseases: the case of diabetes and pre-diabetes. BMC Med Inform Decis Mak 10(1):1–7
35. Polat K, Gunes S, Arslan A (2008) A cascade learning system for classification of diabetes disease: generalized discriminant analysis and least square support vector machine. Expert Syst Appl 34(1):482–487
36. Jabbar MA, Deekshatulu BL, Chandra P (2013) Classification of heart disease using K-nearest neighbor and genetic algorithm. In: International conference of computational intelligence: modeling techniques and applications, pp 85–94
37. Odom MD, Sharda R (1990) A neural network model for bankruptcy prediction. Int Jt Conf Neural Netw 2:163–168
38. Gourisaria MK, Das S, Sharma R, Rautaray SS, Pandey M (2020) A deep learning model for malaria disease detection and analysis using deep convolutional neural networks. Int J Emerg Technol (IJET) 11(2):699–704
39. Syarif I, Prugel-Bennett A, Wills G (2016) SVM parameter optimization using grid search and genetic algorithm to improve classification performance. Telkomnika 14(4):1502
40. Visa S, Ramsay B, Ralescu AL, Van Der Knaap E (2011) Confusion matrix-based feature selection. MAICS 710:120–127
41. Ziarko W (1993) Variable precision rough set model. J Comput Syst Sci 46(1):39–59
42. Davis J, Goadrich M (2006) The relationship between Precision-Recall and ROC curves. In: Proceedings of the 23rd international conference on Machine learning, pp 233–240
43. Chicco D, Jurman G (2020) The advantages of the Matthews correlation coefficient (MCC) over F1 score and accuracy in binary classification evaluation. BMC Genomics 21(1)
44. Sharma R, Gourisaria MK, Rautray SS, Pandey M, Patra SS (2020) ECG Classification using Deep Convolutional Neural Networks and Data Analysis. Int J Adv Trends Comput Sci Eng 9:5788–5795
45. Rautaray SS, Dey S, Pandey M, Gourisaria MK (2020) Nuclei segmentation in cell images using fully convolutional neural networks. Int J Emerg Technol 11:731–737
46. Bharati N, Das S, Gourisaria MK (2021) A review on mobile cloud computing. intelligent and cloud computing springer. Singapore, pp 209–218

Chapter 6
Smartphone Retail-Purchase Buying Intentions: Investigation of Planned Behaviour in Indian Context

Lingam Naveen, Rabi N. Subudhi, Abu Saleh, Aishwarya Mohanty, and Smruti Malhar Mahapatro

1 Introduction

India, in the past decade, has witnessed a tremendous growth in the number of smartphone user. Smartphone has moved from a luxury to necessity during this tenure [1]. A recent research report [2] stated that India is only second to China in terms of smartphone penetration. The report showed that in 2022 India had 647.53 million smartphone users, which account to 45.7% smartphone penetration in the country. From USD 139 billion in 2021, Indian smartphone market is projected to grow at 10.5% CAGR through 2028 to reach USD 281 billion market [3]. This creates a lot of business opportunities for various smartphone marketers around the world to grab a piece of the pie. Multinational brands are investing in India to tap the consumers of rural and semi-urban population.

Consumer options in India's organized retail sector, which is continuously evolving, have been substantially expanded [4]. The surge in e-commerce usage can be highly contributed to expanding smartphone penetration [5]. During the pandemic time, when most academic and professional tasks switched to digital mediums, the significance of smartphones increased. A consumer in India has a variety of places to choose from where to purchase a smartphone. There are now more options than the conventional mobile stores. Smartphone buying-related survey showed that an

L. Naveen (✉)
Biju Patnaik Institute of Information Technology and Management Studies (BIITM), Bhubaneswar, India
e-mail: lingamnaveen3@gmail.com

R. N. Subudhi · A. Mohanty · S. M. Mahapatro
KIIT School of Management, KIIT University, Bhubaneswar, Odisha, India

A. Saleh
Canberra Business School, University of Canberra, Canberra, ACT, Australia
e-mail: abu.Saleh@canberra.edu.au

M. Sahni et al. (eds.), *Novel Developments in Futuristic AI-based Technologies*, Algorithms for Intelligent Systems, https://doi.org/10.1007/978-981-99-3076-0_6

increasing 47% of the Indian buyers prefer to buy smartphones online [6]. Rest 28% still prefer offline and 25% buyers are not sure about their choice of retail format. Thus, exploration of the factors involved in selection of retail format is important from research perspective.

In the process of shopping [7], the choice of the retail format [8] is an important decision taken by consumers. Consumer behaviour is undergoing continuous transformation with the presence of multiple channels [9]. Consumer electronics shops are facing rising competition as a result of the growing variety of retail formats available to customers [10]. Younger shoppers are particularly embracing online purchasing [11]. Understanding various retail format qualities is crucial for studying channel choice behaviour because the internet has significantly changed Indian customers' purchasing habits [12]. As a result, this research concentrated on a single product category, namely smartphones, to allow the effects of various retail models to be seen.

Earlier research works have reinforced those motivational values related to shopping change with the channel of retail [13].

Marketing scholars [14] and practitioners [15] try to understand and study these shopping motivational values and related buying behaviour. It is believed that this study on consumer channel preferences related to smartphone purchase to come up with better retail and marketing strategies [16]. By separating channel users in Odisha, India, this study aims to develop a prediction model on retail format choice (Online or Offline) for smartphone purchase, in light of the paucity of literature on the prediction of shop choice behaviour [17].

2 Review of Literature

2.1 Smartphone Purchase

Product category has an impact on the retail channel choice [12, 13, 18]. Although electronics products are generally categorized as utilitarian products by past literature [19], the current usages of a smartphone make it more than just that. With the fast-growing Indian telecom sector, smartphones have become an essential part in one's personal and business life [20]. The frequency of smartphone purchase has increased overtime. Certain factors were identified as more important than others when customers buy tech-savvy electronic products like smartphone [21].

Salesperson Assistance. Customer service in terms of interpersonal [19, 22] and information assistance [19] at the time of purchase plays an important role in retail format selection especially for technically complex products [10, 23] like smartphones. It not only helps the consumer to make an informed decision but also makes them feel special by catering to their hedonic needs [19]. Customer's loyalty towards an offline store is enhanced by a knowledgeable salesperson [24].

Information Seeking Behaviour. The ease with which information may be accessed, searched for, and combined from a number of sources, particularly when buying speciality electrical devices that require for a deeper comprehension of technicalities. Research indicates that highly informed consumers seek out information to bargain for lower pricing [25].

Customers can get comprehensive product information, recommendations, reviews, video tutorials, and the option to personalize their search experience to their requirements through the internet channel [26]. Customer search and purchase history are stored and retrievable, as well as personalize their screening and selection process, using online shopping platforms [27, 28].

An essential component of the offline retail platform is that it enables clients to get customized, reliable information and assist in personalizing the goods (or services) on spot [26]. Customers can more accurately assess the quality and trustworthiness of a product's various characteristics when they shop in a physical store because they have an instant "sensory or haptic experience" of the actual thing they are considering [27]. Customers' attachment to quality can be strengthened by their direct engagement with the product.

Customer Satisfaction. The retail customers who purchase electronic items online consider the purchase value to be a key factor when measuring satisfaction and are more sensitive to satisfaction when making [29] repurchase decisions. On the other hand, consumer expectations and the overall standard of the purchasing process are the more important factors that influence [30] customer satisfaction with offline purchases.

2.2 *Demographic Differentiators—Online Versus Offline*

Literature suggests that demographic characteristics have an impact on retail format choice behaviour [31–33]. Although several studies have investigated the relationship between demographics and channel choice behaviour, the findings observed were inconsistent [17] and inconclusive.

Gender. The previous studies have shown varied results for consumer behaviour with respect to gender. Studies of [34, 35] found that women have higher inclination for online shopping. Whereas, in contrary, other studies have discussed that men have higher affinity for shopping online [36, 37], especially electronics items [38]. Further, there are studies which found that gender doesn't influence online buying behaviour [4, 39].

Age. Similarly, with respect to age, studies have found various results. Age has negatively [34, 37, 40] and positively [41] impacted online purchase tendencies. However, [4] has asserted that age significantly contributes to retail channel choice decision.

Income, Education and Occupation. In context of income, studies have drawn conclusions related to no impact [42, 34] and positive impact on online shopping manners [41, 43–45]. Educational qualification wise, previous research works have found both significant [40, 44, 45] and non-significant [34, 39] impact on online shopping purchase decision. With respect to occupation or type of work, it has been found that occupation too had significant [46, 47] and non-significant [4] influence on store choice decision.

Thus, it is important to study the impact of socio-demographic factors on retail choice behaviour for smartphone purchase.

2.3 Channel Choice Differentiators—Online Versus Offline

The "bundle of services" that different channels offer, in addition to their products, appeal to different customer groups, depending on their modus operandi [48]. Studies pertaining to consumer behaviour in retail have found links between the type of retail setting and the reasons people shop [49]. Consumers search for various benefits from the channel they shop in depending on the product type [19]. This is due to the complicated character of shopping behaviour [8] in varied circumstances [50].

Price. Price charged against the purchase of a product or service [51] is a significant factor [52], which has an impact on customer purchase intention at the time of buying a smartphone (Tran 2018). Extant literature tends to indicate that the offline prices are higher than online prices [8, 53] and hence consumers choose to purchase from online [54, 55].

Offers and discounts. Consumers prefer online channel for the offers and discounts they get [11]. In Indian online shopping scenario, retailers give offers and discounts to gain more market share [5]. Deep discounts offered by online retails make the competition tough for offline mobile stores [4]. Individuals pick online channel due to an assortment of utilitarian and hedonic factors; out of which rebate and exclusive offers is one of the most significant factors [56].

Service quality. Customer satisfaction is likely to be positively impacted by overall service quality, which is a measure of whether customers' personal needs were addressed (personalization), the trustworthiness of the product (and service) and their overall ratings of the previous consumption experience [57].

Due to the opportunity of interacting with a human salesperson and a pleasant store environment, as well as the perceived decrease in risk associated with online shopping, the customers' perceptions will be more influential in offline purchases than online transactions [26]. Customers those who purchase offline benefit from an immediate "sensory or haptic experience" of the item they are considering, which increases their confidence when making quality judgements [27]. Adversely, shoppers believe that making purchases online is riskier than making purchases offline because of the ambiguity around the dependability of the goods and services. Customers who shop online frequently worry about the product's performance and quality, as well as whether the item will be in fine condition or match the description.

However, when a customer decides to shop in a store, they must travel, park, walk around, look for the products, push their carts, spend time talking with the salespeople about options, and wait in checkout lines all of which result in a variety of transaction costs [58]. Strong evidence suggests that consumers prioritize convenience in their purchasing decisions and choose products or services that can lower their convenience costs. According to a recent study, the biggest convenience-related factor driving people to choose online shopping is the option to make purchases whenever they want, from the comfort of their homes.

Convenience. [59] found that convenience is the most important motivational factor for online purchase due to around-the-clock availability and delivery services. The shoppers who look for saving time [60] and convenience prefer online shopping. Online product information is easier to get than through traditional means, and it allows online buyers to escape pushy salespeople and the bother of packed stores [61].

Ease of Product Comparison. Comparing products and services while shopping online gives customers access to more information [62]. Online channel permits buyers to look for product description, analyse costs and make fast buys.

Internet Usage. People are increasingly using the World Wide Web (WWW) to fulfil a range of informational, communication and entertainment demands as a result of the information revolution's rapid advancement and increased accessibility [63]. People use this robust tool to stay in touch with and grow their social support networks, stay current with international sporting and news events and seek for various forms of information to help them in their daily life.

Peer Influence. Previous studies found that social influence has a greater impact on consumer's purchase intention [52]. In the context of both online and offline buying behaviour, recommendations from family, friends and co-workers have a significant impact on attitude and purchase intention. It can occasionally start to have a bigger impact than the other variables [64]. Family and friend recommendations are more likely than any other source to have an impact on offline consumers [12, 65, 66].

Customer Loyalty. Single-click ordering methods are now widely available from internet shops. According to [67], these characteristics minimize the costs of further purchases, boost client confidence, and finally "lock" the client to the retailer [68]. Because future purchases require less mental load and are more efficient, customers are more likely to return if their earlier interactions with the merchant were positive [69].

On the other hand, because online shopping risk is frequently seen as being higher, customers who have higher expectations for dependability and quality will prefer the relative certainty of offline purchases and reduce the possibility of unfavourable invalidation [29]. For a dependable and risk-free experience, these customers could be more willing to pay the convenience fees (i.e., a trip to the store). In order to keep clients coming back for more purchases, meeting their requirements comes first.

Thus, the determinants of retail format choice should be explored in various contexts [70].

3 Objectives

- To discern between the purchasing advantages sought after by offline and online smartphone buyers.
- To determine and evaluate the motivational elements that affect smartphone purchases.
- To develop a predictive model to anticipate whether smartphone customers will prefer online or offline shopping channels.

4 Methodology

For the exploratory study, a mixed-method approach was adopted. The qualitative approach was considered to triangulate the factors identified from literature review and to generate new factors related to smartphone purchase channel choice behaviour. It was then followed by creating a study instrument for quantitative methodology.

4.1 Qualitative Study

Twenty semi-structured interviews were conducted using an interview guide to understand the process of buying and different factors influencing the retail choice decision. A convenience sampling design was adopted to select the respondents. Out of the 20 respondents, 14 were consumers (online and offline combined) who had purchased a new smartphone within the last 12 months. These respondents were chosen from different genders (10 males and 4 females), age groups (22–50 years old), geographies (4 urban, 7 semi-urban, 3 rural) and occupations. Interviews were also conducted with two mobile retail shop owners (one urban and one semi-urban), three company employees (one regional sales head, one consumer sales trainer and one online marketing associate) working with smartphone brands and one retail store salesperson. These interviews helped in validation and triangulation of selected factors and identification of one new factor, i.e., importance of finance options while buying a smartphone and choosing the retail format, which was included in the study instrument.

4.2 Quantitative Study

The study instrument was designed taking into consideration the factors identified from literature review and semi-structured interviews. Only those respondents who are residents of the state of Odisha and have recently (within a year) purchased a smartphone were requested to participate.

For the logistic regression analysis, 12 shopping motivational variables were created on a five-point Likert scale, and respondents were asked to rate the significance of each item during the purchase of a smartphone on a scale from strongly disagree to strongly agree, from strongly disagree (denoted by 1) to strongly agree (denoted by 5). Depending on the respondents' most recent smartphone purchase channel, the final component of the questionnaire was divided into a separate set of five questions for consumers shopping online and in physical stores. Before gathering information from respondents, the instrument underwent face validity validation by academic specialists. 50 responses were used in pilot research to test for multi-collinearity among the chosen variables.

4.3 Sampling and Data Collection

Collection of data was done through both physical and online mode using convenience sampling, which is a non-probability sampling technique. The wordings of the questionnaire were presented in both languages namely English and regional language (Odia) to overcome barriers in language for rural and semi-urban respondents.

The final sample included 533 replies, of which 273 were from offline consumers and 260 were from online smartphone purchasers. The gathered information was then analysed.

5 Analysis of Data and Findings

5.1 Profiling the Smartphone Consumers

Data were collected from 533 respondents and were analysed for the objectives of the study. Samples were collected from 28 out of 30 districts of Odisha. 243 females and 290 males participated in the survey contributing to 45.6% and 54.45% respectively. Age of the respondents was captured in 4 age brackets i.e., 14–25 years (35.1%), 26–35 years (32.6%), 36–50 years (21.6%) and >50 years (10.7%). Educational qualification and annual family income of the respondents were also captured in the questionnaire and presented in Table 2. Occupation of the respondents was converted to binary variable, i.e., job holders and no job holders for analysis. In order to obtain statistical difference between groups of social-demographic variables and retail format choice, chi-square analysis was performed and the result is summarized in Table 1.

There was no significant difference between groups with respect to educational qualification. However, there was significant difference among the groups with

Table 1 Demographic profiling

Variables	Online		Offline (Retail store)		Total		Chi-square
Retail format choice	Frequency (n = 260)	(%)	Frequency (n = 273)	(%)	Frequency (n = 533)	(%)	
Gender							
Female	99	38	144	53	243	46	11.554[a]
Male	161	62	129	47	290	54	
Total	260	100	273	100	533	100	
Age							
14–25	84	32	103	38	187	35	19.320[a]
26–35	108	42	66	24	174	33	
36–50	45	17	70	26	115	22	
>50	23	9	34	12	57	11	
Total	260	100	273	100	533	100	
Educational qualification							
Primary education	2	1	3	1	5	1	8.471
Matriculation	4	2	6	2	10	2	
Higher secondary	24	9	41	15	65	12	
Diploma	4	2	7	3	11	2	
Graduation	106	41	111	41	217	41	
Post-Graduation	104	40	84	31	188	35	
Doctorate degree	16	6	21	8	37	7	
Total	260	100	273	100	533	100	
Annual family income							
<2 Lakhs	61	23	103	38	164	31	19.657[a]
2–5 Lakhs	52	20	63	23	115	22	
5 Lakhs–10 Lakhs	85	33	71	26	156	29	
5 Lakhs–10 Lakhs	62	24	36	13	98	18	
Total	260	100	273	100	533	100	
Occupation							
Not a job holder	88	34	134	49	222	42	12.724[a]
Job holder	172	66	139	51	311	58	
Total	260	100	273	100	533	100	

Note(s) [a]Significant at 0.05 level

Table 2 Independent variables—binary logistic analysis

Variables in the equation	B	S.E	Wald	df.	Sig.
Age	0.248	0.122	4.248	1	0.021[a]
Gender (1)	−0.414	0.215	2.967	1	0.085
Occupation (1)	−0.453	0.232	4.172	1	0.036[a]
Educational qualification	−0.026	0.097	0.053	1	0.743
Annual family income	−0.511	0.114	15.398	1	0.001[a]
X_1-Importance of price of the smartphone	−0.239	0.124	4.284	1	0.031[a]
X_2-Importance of service quality	0.603	0.123	23.047	1	0.000[a]
X_3-Importance of Salesperson Assistance while buying a smartphone	0.313	0.116	6.044	1	0.026[a]
X_4-Importance of offers and discounts while buying a smartphone	−0.544	0.147	15.121	1	0.000[a]
X_5-Importance of financing options	0.047	0.112	0.226	1	0.695
X_6-Importance of peer influence while buying a smartphone	−0.067	0.105	0.248	1	0.631
X_7-Importance of customer satisfaction while buying a smartphone	0.568	0.152	13.661	1	0.000[a]
X_8-Importance of information seeking while buying a smartphone	−0.492	0.154	9.483	1	0.001[a]
X_9-Importance of customer loyalty to the place of purchase	0.589	0.145	15.446	1	0.000[a]
X_{10}-Importance of convenience while buying a smartphone	−0.369	0.169	4.292	1	0.041[a]
X_{11}-Importance of ease of product comparison while buying a smartphone	−0.306	0.156	4.549	1	0.031[a]
X_{12}-Importance of internet usage in general day to day life	−0.062	0.147	0.124	1	0.712
Constant	1.312	0.645	5.933	1	0.011[a]
−2 Log likelihood	532.272	($P < 0.001$)			
Cox and Snell R square	0.286				
Nagelkerke R square	0.343				

Note(s) [a]Significant at 0.05 level

respect to gender ($\chi^2 = 11.554$), age ($\chi^2 = 19.320$), annual family income ($\chi^2 = 19.657$) and occupation ($\chi^2 = 12.724$).

90.1% of the respondents purchased smartphone for themselves and rest (9.9%) purchased for others (their relative, friends and colleagues). And there is no significant difference between online and offline purchase groups ($\chi^2 = 1.975$, $p > 0.05$). Although, 77.1% of the respondents bought one smartphone in last 1 year and rest 22.9% had bought more than one smartphone in last 1 year. But statistically, there is no significant difference between online and offline smartphone buyers ($\chi^2 = 0.263,\ p > 0.05$).

5.2 *Predictive Model Development—Online Versus Offline Smartphone Consumers*

To develop a model that predicts retail channel choice of a consumer while buying a smartphone, binary logistic regression was used [71]. Based on the selected shopping motivation triggers and socio-demographic variables chosen, the model intends to classify smartphone buyers as an online or offline buyer.

Hence, the dependent variable is the place of purchase:

0—Online consumers.

1—Offline (Retail store) consumers.

The independent variables were tested for reliability and multi-collinearity. The results showed that the scale was reliable (Cronbach's $\alpha = 0.925$). The values of VIF and tolerance levels were well within acceptable range indicating that multi-collinearity was negligible and the independent variables could be used for binary logistic regression.

The expression for logistic regression is:

$$ln\left(\frac{P}{1-P}\right) = \alpha_0 + \alpha_1 X_1 + \alpha_2 X_2 + \cdots + \alpha_k X_k \tag{1}$$

where P is the probability of the consumer being classified as offline smartphone shopper and (1–P) is the probability of the shopper being an online consumer. The socio-demographic variables used for regression are gender (G), age (A), educational qualification (E), annual family income (I) and occupation (OP). Based on Eq. (1), along with these socio-demographic variables, 12 shopping motivation factors were taken (X_1–X_{12}) into consideration to generate the following logistic regression equation:

$$\begin{aligned} ln\left(\frac{P}{1-P}\right) &= 1.312 - 0.414G + 0.248A - 0.026E - 0.511I \\ &- 0.453OP - 0.239X_1 + 0.603X_2 + 0.313X_3 - 0.544X_4 \\ &+ 0.047X_5 - 0.067X_6 + 0.568X_7 - 0.492X_8 + 0.589X_9 \\ &- 0.369X_{10} - 0.306X_{11} - 0.062X_{12} \end{aligned} \tag{2}$$

Statistical values related to each variable are reported in Table 2.

Based on the above logistic regression equation, if the calculated P value of a particular respondent is >0.5, then the consumer is more likely to be buying his/her smartphone from a physical retail store (offline). And, if the value of probability (P) is found to be <0.5, then the respondent is more likely to be an online consumer.

The goodness-of-fit of the model was tested using Omnibus test and Hosmer & Lemeshow test statistics. Both the tests indicated that the model has a good fit.

Among the socio-demographic variables, gender and educational qualification were not found to be significantly impacting the retail format choice decision. Other variables like age (B = 0.248, Exp. B = 1.281), occupation (B = −0.453, Exp. B = 0.636) and family income (B = −0.511, Exp. B = 0.599) were found to be significant predictors in the model. A positive coefficient for age suggests that with increasing age, customers are more likely to purchase smartphones through offline outlets. In case of occupation, the coefficient is negative suggesting that a non-job holder is more likely to shop for smartphone online by ($e^{-0.453}$) 0.636 times than a job holder. Similarly, as the annual family income increases the likelihood of purchasing smartphone online increases.

X5 (financial alternatives), X6 (peer influence) and X12 (internet usage) were found to have no significant impact on predicting the choice of retail format in an analysis of shopping incentive factors. In the model, it was discovered that X2 (service quality), X3 (salesperson assistance), X7 (customer satisfaction) and X9 (customer loyalty) were positive and significant predictors. This suggests that when these components increase by one unit, the odd ratio in favour of offline retail purchase goes up by ($e^{0.603}$) 1.828, ($e^{0.313}$) 1.368, ($e^{0.568}$) 1.765 and ($e^{0.589}$) 1.802 units, respectively. Contrarily, X_1 (price), X_4 (offers and discounts), X_8 (information seeking), X_{10} (convenience) and X_{11} (ease of product comparison) were found to have negative coefficient in the predictive model suggesting that one-unit increase in these factors increase the odd ratio in favour of online smartphone purchase by ($e^{-0.239}$) 0.787, ($e^{-0.544}$) 0.580, ($e^{-0.492}$) 0.611, ($e^{-0.369}$) 0.691 and ($e^{-0.306}$) 0.736 units, respectively.

The model could correctly classify 74.9% of the observations. 73.8% of online consumers and 75.9% of offline consumers were classified correctly by the predictive model. With significant goodness-of-fit, predictive capabilities and chi-square statistics, the model can be considered as a reliable predictive model.

5.3 *Analysis of Online Consumers*

The responses from online consumers were separately assessed by a dedicated section in the instrument. 75.4% of the online respondents would prefer buying their next smartphone online. Only 2.7% would like to switch their channel to retail store for their next smartphone purchase and 21.9% respondents are not sure about their next purchase channel. Out of 260 online shoppers of smartphone, only 19 respondents (7.3%) responded by saying that they would have bought their phone from retail store if there was no lockdowns and shutdowns due to COVID-19 pandemic. Few

Table 3 Online consumers: n = 260

One-sample statistics	Mean	T
I look for customer review online and it is important to me before purchasing a smartphone	4.485	77.594[a]
The product exchange and return facility offered by online platforms is important for me while buying my smartphone	4.542	83.656[a]
I wait for online special discount days/sales offers (e.g., Big Billion Day, Great Indian Festive Sale etc.) to buy my smartphone	4.081	55.866[a]
I like the overall shopping experience online when I purchase from online platforms	4.404	85.996[a]

Note(s) [a]Significant at 0.05 level

statements related to online smartphone purchase driver were asked to respondents and assessed using one-sample t-test to find out their importance to online consumers (Table 3). The important factors identified were online reviews, return and exchange facilities provided by online retailers, special discount days or sales offers like Big Billion Day and the overall online shopping experience.

5.4 Analysis of Offline (Retail Store) Consumers

When offline buyers were assessed separately, it was found that only 54.6% of the respondents would prefer to buy their next smartphone from a retail store. 15.4% would switch to online channel and 30.0% are not sure about the choice of channel for their next purchase. Responses of five statements related to offline smartphone purchase were collected from retail store buyers exclusively and analysed using one-sample t-test to find out their importance (Table 4). The test results indicated that all the factors were important for offline smartphone buyers, i.e., researching online before purchase, prompt service of the retailers, personally knowing the retail store owner/being a regular customer, better bargaining power and overall retail store shopping experience.

6 Discussion and Conclusion

This study aimed to identify and evaluate several demographic and purchasing trigger elements that affected consumers' decisions to purchase smartphones from offline or online retailers. It was discovered that the motivating elements of various retail channels differ significantly from one another. Based on the demographic and purchasing motivational characteristics, the study produced a predictive model

Table 4 Offline consumers: n = 273

One-sample statistics	Mean	T
I generally research online about smartphones before purchasing from store	4.366	67.213[a]
I like the prompt service retail store owners and store sales personnel provide	4.231	67.880[a]
I know the mobile store owner/salesperson personally and I am a regular customer of the	3.117	33.729[a]
Buying from retail store (offline) gives me a better bargaining power compared to online purchase	3.418	38.590[a]
I like the overall shopping experience when I purchase from retail store	4.242	68.583[a]

Note(s) [a]Significant at 0.05 level

to forecast the retail format choice of customers. Using binary logistic regression modelling, some socio-demographic antecedents were also discovered to significantly contribute to the predictive model.

7 Study Implications

This study also identified the key elements, based on a thorough review of the literature, that must be taken into account when predicting consumers' behavioural intentions, which have an impact on their choice of retail channel. This paper attempts to add important bits of knowledge to retail market by predicting different trigger points of consumer while purchasing smartphone from different format choice like online and offline stores.

A mixed method approach was used to identify 12 characteristics, and a predictive model with a classification accuracy of more than 74% was proposed to predict the retail format choice for smartphone purchase. Using binary logistic regression modelling, some socio-demographic antecedents were also discovered to significantly contribute to the predictive model. Chi-square analysis was also utilized to distinguish between other behavioural factors associated with smartphone purchase. The ability to compare items easily, offers and discounts, price, information-seeking behaviour and active internet usage were discovered to be differentiators for online smartphone purchases. In contrast, it was found that offline purchases were more dependent on retail service quality, salesperson assistance, customer satisfaction and customer loyalty.

Due to the similarity of retail market conditions and consumer purchasing patterns, the study's findings about smartphone purchases can, therefore, be generalized to the Indian consumer, making information on purchasing reasons in relation to retail format choice behaviour important and helpful for the smartphone companies and retailers when establishing plans for channel variation and growth.

Our findings largely support the suggested retail strategies for enhancing the customer experience, such as enabling customers to interact with best-quality and reliable products in a pleasant atmosphere and giving them access to competent salespersons. This study will help the retailers who want to boost their sales as it is much more essential to increase offline sales by running marketing efforts, managing word-of-mouth and offering accurate product information to better control [29] pre-purchase expectations. Since the distinctive qualities of the offline channel are difficult to recreate in the online environment and since customers look for better purchase value when they shop online, offline businesses might gain an advantage over their online rivals by offering competitive prices.

The online retailers, on the other hand, require efficient websites that provide cheaper searches, quicker and more secure checkouts, simple access to previous order and search history, and straightforward mobility with independent customization. To aid in memory retention and recall, it can be beneficial to focus on the information accessibility [29] and content richness by including product comparisons and instructional videos on the product page.

8 Limitations and Future Research

This study examined a number of parameters and included them in the predictive model, but it's possible that there are still other variables that need to be investigated. This study's domain was restricted to one Indian State-Odisha. For generalizability, comparable research can be carried out across geographical regions, different product categories, other demographic variables and service industry as well.

References

1. Watkins J, Kitner KR, Mehta D (2012) Mobile and smartphone use in urban and rural India. Continuum 26(5):685–697
2. Top Countries/Markets by Smartphone Penetration and Users | Newzoo. https://newzoo.com/insights/rankings/top-countries-by-smartphone-penetration-and-users. Last accessed 29 Jan 2023
3. India Smartphone Market Size. https://www.researchandmarkets.com/reports/5543303/india-smartphone-market-size-analysisgrowth?utm_source=CI&utm_medium=PressRelease&utm_code=dqn487&utm_campaign=1674100+-+India+%24139+Billion+Smartphone+Markets+to+2028&utm_exec=chdo54prd. Last accessed 28 Jan 2023
4. Gupta AK, Shukla AV (2015) Store choice behaviour for consumer durables in NCT-Delhi: effect of shopper's demographics. Paradigm 19(2):152–169
5. Richa H, Vadera S (2019) Determinants of online shopping behaviour in India. Int J Recent Technol Eng 8(3):3946–3950
6. 91mobiles Smartphone Buyer Insights Survey 2020: a summary, https://www.91mobiles.com/hub/91mobiles-smartphone-buyer-insights-survey-2020/. Last accessed 8 Aug 2022
7. Peterson RA, Balasubramanian S, Bronnenberg BJ (1997) Exploring the implications of the Internet for consumer marketing. J Acad Mark Sci 25:329–346
8. Hsiao MH (2009) Shopping mode choice: physical store shopping versus e-shopping. Transp Res Part E: Logist Transp Rev 45(1):86–95

9. Van Bruggen GH, Antia KD, Jap SD, Reinartz WJ, Pallas F (2010) Managing marketing channel multiplicity. J Serv Res 13(3):331–340
10. Carpenter JM, Balija V (2010) Retail format choice in the US consumer electronics market. Int J Retail Distrib Manag 38(4):258–274
11. Sarkar R, Das S (2017) Online shopping versus Offline shopping: a comparative study. Int J Sci Res Sci Technol 3(1):424–431
12. Prasad J, Raghu Y (2018) Determinant attributes of online grocery shopping in india-an empirical analysis. IOSR J Bus Manag 20(2):18–31
13. Wong RM, Wong SC, Ke GN (2018) Exploring online and offline shopping motivational values in Malaysia. Asia Pac J Mark Logist 30(2):352–379
14. Mokhlis S, Salleh HS (2009) Decision-making styles of young Malay, Chinese and Indian consumers in Malaysia. Asian Soc Sci 5(12):50–59
15. Hong IB (2015) Understanding the consumer's online merchant selection process: the roles of product involvement, perceived risk, and trust expectation. Int J Inf Manag 35(3):322–336
16. Rangaswamy A, Van Bruggen GH (2005) Opportunities and challenges in multichannel marketing: an introduction to the special issue. J Interact Mark 19(2):5–11
17. Choi J, Park J (2006) Multichannel retailing in Korea: effects of shopping orientations and information seeking patterns on channel choice behavior. Int J Retail Distrib Manag
18. Balasubramanian S, Raghunathan R, Mahajan V (2005) Consumers in a multichannel environment: product utility, process utility, and channel choice. J Interact Mark 19(2):12–30
19. Haridasan AC, Fernando AG (2018) Online or in-store: unravelling consumer's channel choice motives. J Res Interact Mark
20. Mohan A (2014) Consumer behaviour towards smartphone industry in Indian market (Doctoral dissertation, Dublin Business School)
21. Mohan Raj P, AK M (2015) Psychographic profiling of Indian young adult consumers of smartphone-VALS approach. Manag Stud Econ Syst (MSES) 2(2):95–102
22. Baker J, Parasuraman A, Grewal D, Voss GB (2002) The influence of multiple store environment cues on perceived merchandise value and patronage intentions. J Mark 66(2):120–141
23. Bencic E (2017) David Jones to sell electronics again
24. Rabbanee FK, Ramaseshan B, Wu C, Vinden A (2012) Effects of store loyalty on shopping mall loyalty. J Retail Consum Serv 19(3):271–278
25. Chiou-Wei SZ, Inman JJ (2008) Do shoppers like electronic coupons? a panel data analysis. J Retail 84(3):297–307
26. Laroche M, Yang Z, McDougall GH, Bergeron J (2005) Internet versus bricks-and-mortar retailers: an investigation into intangibility and its consequences. J Retail 81(4):251–267
27. Degeratu AM, Rangaswamy A, Wu J (2000) Consumer choice behavior in online and traditional supermarkets: the effects of brand name, price, and other search attributes. Int J Res Mark 17(1):55–78
28. Shankar V, Smith AK, Rangaswamy A (2003) Customer satisfaction and loyalty in online and offline environments. Int J Res Mark 20(2):153–175
29. Hult GT, Sharma PN, Morgeson III FV, Zhang Y (2019) Antecedents and consequences of customer satisfaction: do they differ across online and offline purchases? J Retail 95(1):10–23
30. Cotten SR, Gupta SS (2004) Characteristics of online and offline health information seekers and factors that discriminate between them. Soc Sci Med 59(9):1795–1806
31. Gillett PL (1976) In-home shoppers—an overview: this article reviews findings on in-home shoppers and points out further research needs. J Mark 40(4):81–88
32. Jasper CR, Lan PN (1992) Apparel catalog patronage: Demographic, lifestyle and motivational factors. Psychol Mark 9(4):275–296
33. Wright G (1996) Aging boomers will revamp retail scene. Brandweek 37(39):18
34. Ding Y, Lu H (2017) The interactions between online shopping and personal activity travel behavior: an analysis with a GPS-based activity travel diary. Transportation 44:311–324
35. Maat K, Konings R (2018) Accessibility or innovation? Store shopping trips versus online shopping. Transp Res Rec 2672(50):1

36. Akhter SH (2003) Digital divide and purchase intention: why demographic psychology matters. J Econ Psychol 24(3):321–327
37. Crocco F, Eboli L, Mazzulla G (2013) Individual attitudes and shopping mode characteristics affecting the use of e-shopping and related travel. Transp Telecommun 14(1):45
38. Zhen F, Cao X, Mokhtarian PL, Xi G (2016) Associations between online purchasing and store purchasing for four types of products in Nanjing, China. Transp Res Record 2566(1):93–101
39. Beldona S, Racherla P, Mundhra GD (2011) To buy or not to buy: Indian consumers' choice of online versus offline channels for air travel purchase. J Hosp Mark Manag 20(8):831–854
40. Zhou Y, Wang XC (2014) Explore the relationship between online shopping and shopping trips: an analysis with the 2009 NHTS data. Transp Res Part A: Policy Pract 70:1–9
41. Donthu N, Garcia A (1999) The internet shopper. J Advert Res 39(3):52
42. Farag S, Weltevreden J, Van Rietbergen T, Dijst M, van Oort F (2006) E-shopping in the Netherlands: Does geography matter? Environ Plan B: Plan Des 33(1):59–74
43. Farag S, Schwanen T, Dijst M, Faber J (2007) Shopping online and/or in-store? a structural equation model of the relationships between e-shopping and in-store shoppingTransp Res Part A: Policy Pract 41(2):125–141
44. De Blasio G (2008) Urban–rural differences in internet usage, e-commerce, and e-banking: Evidence from Italy. Growth Chang 39(2):341–367
45. Cao XJ, Xu Z, Douma F (2012) The interactions between e-shopping and traditional in-store shopping: an application of structural equations model. Transportation 39:957–974
46. Ren F, Kwan MP (2009) The impact of geographic context on e-shopping behavior. Environ Plan B: Plan Des 36(2):262–278
47. Cao X, Chen Q, Choo S (2013) Geographic distribution of e-shopping: application of structural equation models in the Twin Cities of Minnesota. Transp Res Rec 2383(1):18–26
48. Furey T, Friedman L (2012) Channel advantage. The Routledge
49. Dawson S, Bloch PH, Ridgway NM (2002) Shopping motives, emotional states, and retail outcomes. The Environments of Retailing. Routledge, London, pp 65–81
50. Dellaert BG, Arentze TA, Timmermans HJ (2008) Shopping context and consumers' mental representation of complex shopping trip decision problems. J Retail 84(2):219–232
51. Kotler P, Armstrong GM (2010) Principles of marketing. Pearson Education India
52. Lin NH, Lin BS (2007) The effect of brand image and product knowledge on purchase intention moderated by price discount. J Int Manag Stud 2(2):121–132
53. Rotem-Mindali O, Weltevreden JW (2013) Transport effects of e-commerce: what can be learned after years of research? Transportation 40:867–885
54. Jasrotia SS, Mishra HG, Koul S (2019) Brick or click? Channel choice disruptions in travel industry. Asia-Pac J Manag Res Innov 15(1–2):16–26
55. Koyuncu C, Bhattacharya G (2004) The impacts of quickness, price, payment risk, and delivery issues on on-line shopping. J Socio-Econ 33(2):241–251
56. Liu M, Xiao Y (2008) Reach on relationship between consumer satisfaction and web consumer shopping behavior. Int J Bus Res 8(5):127–135
57. Fornell C, Johnson MD, Anderson EW, Cha J, Bryant BE (1996) The American customer satisfaction index: nature, purpose, and findings. J Market 60(4):7–18
58. Chintagunta PK, Chu J, Cebollada J (2012) Quantifying transaction costs in online/off-line grocery channel choice. Market Sci 31(1):96–114 (2012)
59. Robinson H, Dall'Olmo Riley F, Rettie R, Rolls-Willson G (2007) The role of situational variables in online grocery shopping in the UK. Mark Rev 7(1):89–106
60. Akaah IP, Korgaonkar PK, Lund D (1995) Direct marketing attitudes. J Bus Res 34(3):211–219
61. Grewal D, Iyer GR, Krishnan R, Sharma A (2003) The internet and the price–value–loyalty chain. J Bus Res 56(5):391–398
62. Huang MH (2000) Information load: its relationship to online exploratory and shopping behavior. Int J Inf Manag 20(5):337–347
63. Jadad AR, Gagliardi A (1998) Rating health information on the internet: navigating to knowledge or to Babel? Jama 279(8):611–614

64. Román S, Cuestas PJ (2008) The perceptions of consumers regarding online retailers' ethics and their relationship with consumers' general internet expertise and word of mouth: a preliminary analysis. J Bus Ethics 83:641–656
65. Nelson MR, McLeod LE (2005) Adolescent brand consciousness and product placements: awareness, liking and perceived effects on self and others. Int J Consum Stud 29(6):515–528
66. Kulkarni G, Ratchford BT, Kannan PK (2012) The impact of online and offline information sources on automobile choice behavior. J Interact Mark 26(3):167–175
67. Lynch Jr JG, Ariely D (2000) Wine online: search costs affect competition on price, quality, and distribution. Mark Sci 19(1):83–103
68. Johnson EJ, Bellman S, Lohse GL (2003) Cognitive lock-in and the power law of practice. J Market 67(2):62–75
69. Van Birgelen M, De Jong A, De Ruyter K (2006) Multi-channel service retailing: The effects of channel performance satisfaction on behavioral intentions. J Retail 82(4):367–377
70. Shi K, De Vos J, Yang Y, Witlox F (2019) Does e-shopping replace shopping trips? Empirical evidence from Chengdu, China. Transp Res Part A: Policy Pract 122:21–33 (2019)
71. Ben-Akiva ME, Lerman SR, Lerman SR (1985) Discrete choice analysis: theory and application to travel demand. MIT Press

Chapter 7
Automated Breast Cancer Detection Systems (Analytical Study)

Arwa H. Zabian

1 Introduction

Breast cancer is considered one of the predominant cancers in the women; it starts in the breast tissues and spread to the other body tissues, the fifth reason for death in the world.

Breast cancer affects women more than men, but in women, its symptoms appear early and that may reduce the mortality rate. While, in males, the outcomes tend to be poor due to late diagnosis.

Statistics approved on 01/2020 states that in the United States this year 325.010 women were diagnosed with breast cancer, and 2,620 men [32]. Tumor cells are like body cells that make it difficult to be detected by the immune system. Tumor cells are formed of normal cells due to modification or mutation in the DNA or RNA, these modifications or mutations are caused by different factors like radiation in general, viruses, bacteria, fungi, parasites, all these motivations may cause entropy disease that can't be corrected by the organism but it needs external intervention. This entropy disease could eventually convert to cancer. The amount of tumor cells is increased in a manner that the immune system can't eliminate the non-useful cells. The breast consists of two types of tissues, glandular tissues and stromal tissues. Glandular tissues in the breast house the milk, and the stromal tissues of the breast house the fatty and fibrous connective. The lymphatic tissues in the breast are considered an immune system that defends against infection and cancer.

Causes of breast cancer can be genetic, family history, hormonal causes, lifestyle, dietary causes and environmental causes. Management of breast cancer can be by surgery, radiation therapy, chemotherapy, gene therapy. Breast cancer can be benign or malignant.

A. H. Zabian (✉)
Faculty of Science and Information Technology, Jadara University, Irbid, Jordan
e-mail: Arwa@jadara.edu.jo

M. Sahni et al. (eds.), *Novel Developments in Futuristic AI-based Technologies*, Algorithms for Intelligent Systems, https://doi.org/10.1007/978-981-99-3076-0_7

Benign breast cancer (benign breast tumor) that present as a change within the breast tissue including cysts formation (accumulating packets of fluid), fibrosis, lumpiness, and area of thickening, tenderness, or breast pain.

Malignant breast cancer starts in the cells that line the ducts or that line the lobules and may spread to the other surrounding tissues; it is evaluated and classified according to its severity in two types: non-invasive breast cancer and invasive breast cancer.

1. Non-invasive breast cancer is limited to the milk channels and does not extend to the neighboring breast tissues that consist of fat and connective tissue the most frequently occurring type is Ductal Carcinoma in Situ (DCIS), which can reach 90%. While in Lobular Carcinoma in Situ (LCIS), the number of cells is increased significantly within the milk glands (15–20%) [35].
2. Invasive Breast Cancer: Cells penetrate the duct and lobular walls and expand in the fatty and connective tissues of the breast. Invasive Lobular Carcinoma (ILC) starts in the lactiferous glands (lobules) and spreads to the other regions of the body, its percentage from 10 to 15% of invasive breast cancer. Infiltrating Ductal Carcinoma (IDC) begins in the milk channels and breaks the wall of the duct, invading the fatty tissues and the other region of the body; it represents 80% of invasive breast cancer cases [35].

1.1 Breast Cancer Levels

When a cancer is diagnosed, a stage is assigned to it, based on how advanced it is. The stages help doctors in determining the most appropriate treatment and the prognosis. Breast cancer stages can be classified into five stages:

level 0: the tumor is restricted to a milk duct and not invaded the surrounding breast tissue.

Stages from 1 to 4, all are for invasive cancer but differ based on the tumor diameter, and the cells to which is spread.

level 1: the tumor size is less than 2 cm in diameter and limited to this space in the breast, or the tumor is less than 2 cm and expands to one-three lymph nodes in the armpit.

level 2: the tumor diameter is between 1.9 cm and 5 cm and has expanded to one-three lymph nodes in the armpit, or the tumor is larger than 5 cm in diameter but has limited to this space in the breast.

level 3: the tumor diameter is 5 cm or less but has spread to 4–9 lymph nodes, or the tumor is larger than 5 cm and has spread to up to nine lymph nodes, or can be in any size but spread to 10 or more lymph nodes or extended to the chest area wall or skin.

level 4: is called metastasis cancer the tumor has spread to distant organs or tissues such as the lungs, bones [2].

2 Early Breast Cancer Detection

Death rate from breast cancer can be reduced by both improvement in treatment and early detection. The surviving rate depends on the stage at which it is detected. The relative survival rate for 99% of localized disease, 85% of regional disease and 27% of distant stage disease is 5 years. Survival with each stage varies by tumor size, for example, among women with regional disease, the 5-year relative survival is 95% for tumor less than or equal to 2.0 cm, 85% for tumor 2.1–5.0 cm and 72% for tumors greater than 5.0 cm. To decrease the death rate, increased understanding about breast cancer and early discovery remain the most crucial elements. The mortality rate stays diminished in most high-income countries. This rate is higher in developing countries, where breast cancer is the most popular cause of death among adult women in the developing countries [5] because most of the cases in the developing countries are detected in late stage. In the USA in the last years, the incidence of breast cancer is increased due to changes in lifestyle, however, the mortality ratio is decreased due to the introduction of mammography and chemotherapy and other technologies that helped in early detection of breast cancer without using surgical solution. Giving priority to early detection, in developing nations, can lead to a significant increase in the survival rate. Awareness about breast cancer and education efforts should face the reality that many women with less income and education may not seek care and could be unaware of what represents a breast mass [6].

In [7], a survey of the frequency of breast cancer cases in four cancer registries based on populations in the Middle East. (Izmir, Turkey, Cyprus, Jordan, Israel), to determine the factors linked to differences in the rate of breast cancer in different regions, the survey presented during the years 2005–2010, the results presented in the survey are compared with that from the USA center. Dataset compiled by the participants included all cases of in situ and invasive cancer, results showed that the proportion of cases with invasive cancer appeared to rise with age among Jordanian and Turkish participants (higher in Jordanian). However, for Invasive Ductal Carcinoma, the majority of cases in most of the countries involved, in all age groups can be greatly affected by regional differences in mammography utilization, especially when it comes to early-stage cancer. In 2017, according to American Cancer Society, approximately 40.610 women and 640 men are expected to die from breast cancer [8], different factors influence in the survival rate: environmental risk factors, reproductive factors, obesity and diet, age, race, ethnicity, family history and genetic factors. In 2018, the distribution of the death rate distribution in women given the age is about 0.9 from age 20–34 increase to 22.4 from age 65–74 and decreases to 16.9 for age more than 84. Breast cancer mortality is reduced by 40% for women of 40–74 years of age who participate in regular mammogram examination every 1–2 years [9]. As statistics results indicate that there were over 2 million new cases of breast cancer in 2018, the country with the highest rate between women is Belgium with 113.2% for each 100.000, the study indicates that there is strong relationship between alcoholic drinks consumption and the probability of premenopausal breast cancer, and there is a strong relationship between physical activity and protecting against pre-menopausal breast cancer [15]. Clinically breast examination combined with mammography is considered essential to reducing mortality from breast cancer. Mammography is used as a primary tool for detection breast cancer but it cannot detect all types of breast cancer, for that, there is a need to use additional resources for improving the breast cancer prediction or to identify abnormal masses in mammography images. In recent years, digital mammography has also been used as an alternative to traditional mammography. Computer Aided Detection (CAD) can be incorporated into the workstation, and the results of analysis will help the radiologists in detecting suspect lesions.

CAD has been assessed in several studies, which suggest an incremental value in terms of sensibility [3, 4]. Breast cancer affects one in eight women in their life time, the introduction of information technology in the field of medicine reveals a new diagnosis mechanism called medical image processing that is not limited to cancer but can help in diagnosis and treatment the results analysis of other diseases. Medical image processing is a technique that helps in detecting or distinguishing between breast cancers and infecting breast. Early detection helps in proper diagnosis and makes treatment results more effective and minimizes the risk of death. Medical images are different in nature and need special processing or sometime a good experience before diagnosis is made. Most of the medical images are diagnosed by medicine using eye detection, less expert medicine sometimes can mix between cancer and calcification of the breast due to similarities in the appearance. Breast calcification is a small deposit of calcium developed in a women breast tissue. Breast calcification is in two types are: macro calcification that is presented as sparse white dots are dispersed randomly with the breast and are considered non-cancerous. Micro-calcification white dots appear in certain pattern and are clustered together can be benign, probably benign or suspicious [10]. Using some technical tools like artificial intelligence tools can help medicine in distinguishing between breast cancer and calcification, can give a preliminary report about the state of the cancer benign, malign or metastasis cancer. In this paper, we are interested in examining some of the artificial intelligence tools used in mammogram image processing to identify all kind of mass or any abnormality found in the image. Obviously, the detection of abnormal masses in the image cannot be definitely correct or decisive but help some medicine with less experience in predicting or early detection of breast cancer before deciding biopsy. Breast cancer prediction and detection based on artificial intelligence techniques (AI) has attracted the attention of many researchers where is used the basics of image processing (filtering, extracting background, segmentation, feature extraction) to find the important features in the mammogram and then is used AI tools like machine learning, fuzzy system, genetic algorithm, neural networks, decision tree for classification and prediction the abnormality in the breast cells. In this paper, we will give a brief description of some of the AI algorithms used for early breast cancer detection and the prediction accuracy rate of each one. A comparison between presented works in term of accuracy rate is presented.

Processing images to extract the needed information require high computational capabilities. Automated system helps in image processing, feature extraction and analysis of digital mammogram. It is noted that about 10–30% of breast disease is discovered during routine screening [11]. Reading mammography by radiologists may improve the analysis sensibility but increase the cost of screening process [12], thus computer-aided detection (CAD) can act as second reader where the final decision will be made by the radiologists. Image processing passes in the following phases:

Preprocessing phase: in this phase, image filtering is used to eliminate the noise in the image, the noise may come during capturing from different factors like (brightness, variety of color intensity, bad sensor quality, movement during the capturing, etc.) [13].

Segmentation phase: in this phase, the important area in the image is identified, area means set of connected pixels that have the characteristics different from the other pixels of the image.

Feature extraction: after defining the important area in the image, a set of features must be extracted from these areas to construct useful information needed in the image classification, some of these features are:

Perimeter (P): is defined as the number of pixels that surround the selected area

1. Area means part of the image composed on connected pixels similar to each other and different from the other pixel of the image.
2. Compactness $= (P)^2/4\pi$.
3. Area center: is a pixel located at equal distance (near equal) from the pixels that are on the perimeter (border).
4. Diameter is the distance between the central pixel and the pixels that are on the border of the area.
5. Density: is the number of pixels in unit measure.
6. Orientation: defined as the angle of the axis of the selected area with respect to x–y axis.
7. Direction: is defined as the distribution of pixels (left–right) relatively to the center of the area.

Classification phase: all the information collected in the previous phase must be classified and normalized to be used as input to machine learning tools to predict the class of cancer that can be benign or malign. For malign cases, another classification and processing must be performed to define the non-invasive type based on other features like size and shape. In size, we mean the number of pixels on the diameters. The distribution of pixels in the selected area can be in regular shape or non-regular shape [14].

The learning process consists on preparing a machine that is able to predict the future cases based on historical cases. Observable pattern in the image will give information about the image, the accuracy of the classification depends on the feature extraction stage, features are used to predicate relevant pattern in the data that help in solving computational problems. In the next section, we will give a brief description of the most used algorithms in breast cancer detection and the related work in each field. Using automated tools in the medical field have made possible to collect large volume of data, analyzing it and predicting the outcome of the disease in a good accuracy rate.

3 Artificial Intelligence Tools Used in Predicting and Classification Breast Cancer

3.1 Decision Tree Algorithm

Decision tree algorithm is a classifier that partitions the instance space recursively to create a predictive model that can be applied on the observed nodes to give a conclusion about a target value [16]. In the literature, there are different algorithms that are based on decision tree technique used to classify data like ID3 (Iterative Dichotomiser3), C4.5, CS, J48 all these algorithms are extension of Algorithm ID3. [17] proposed automated breast cancer detection mechanism that is based on J48 [18], which is applied on data set using WEKA (Java toolkit) [19], after dividing the data into benign and malignant it is calculated IG (information Gain) for choosing an attribute as a split point to construct the decision tree with leaf nodes as the classes labels (benign, malignant). The attribute of higher information gain on the data used where uniformity of cell size. The results show that 94.56% of cases analyzed (699 cases) were tested correctly (are benign and the system results match (66.2%) or are malignant and the system results match (33.7%), false positive results were 0.02 and false negative were 0.02 that correspond to a percentage of 5.4% for all wrong results), Table 1. In [20], is used a machine learning algorithms are used to predict the cancer is benign or malignant, where is used decision tree and artificial neural networks to create a machine that predict the results. In decision tree machine that uses J48 algorithm, the leaf nodes are divided into two categories benign or malignant. Data are split into 80% training set and 20% test set, then 10 attributes are used to construct the decision tree, in the root, there is clump thickness and, in the last level, there are mitoses that divide the data then to two classes. As results show, the algorithm successfully predicts 82 cases from 140 as benign and 48 malignant cases with accuracy 92.857%, Table 1, then it is implemented ANN to evaluate the results obtained by decision tree algorithm, the input nodes where the 10 attributes used to construct the decision tree, the output of the hidden layer is considered as the input of the two output units that correspond to the results of the diagnosis (benign or malignant) different number of hidden layers were used to obtain best results (9, 2, 2), the results show that 674 cases of 699 were predicted correctly (441 benign, 233 malign) that correspond to accuracy rate 96.42% and 25 cases from 699 were classified wrong that correspond to 3.57% accuracy rate, Table 1.

Table 1 A comparison between the results tested correctly using ID3, J48 and ANN

	Decision tree		ANN
	ID3 [17]	J48 [20]	ANN [20]
Tested correctly benign	66.2	58.5	63.0
Tested correctly malign	33.7	34.2	33.3
Accuracy rate	94.56	92.87	96.42

3.2 ANN: Artificial Neural Networks

An artificial neural network (ANN) can be defined as a model of reasoning based on human brain. Typical ANN model is composed on set of hierarchical layers are: input layers, hidden layers and output layers, each input layer is connected to one or more intermediate layer (hidden) based on well-defined parameters. The number of input layer can be smaller than, or equal, or larger than the hidden layer, the output layers must be smaller than the hidden layer (vary from 1 to 4), the output layer is the classification results. Input layer receives the data, transmits it to the hidden layers that process data to provide the training results to the output layer, the output layer shows the classification results, the performance of machine and the accuracy of the classification results depend on the number of hidden layers and the size of the data [21, 22] presented a classification technique that is based on ANN, in which 70% of data are used for training and 15% of data are used for validation. In the test set, the prediction was 100% corrected. That means, in the training set, the accuracy was 99.1%, in the validation and testing set the accuracy were 100% for a total 99.4% overall of the cases (322 mammogram, 270 are normal and 52 are cancerous). Medical diagnosis is an important task that should be performed in accurate and efficient manner, unfortunately not all doctors have the same skills and experience and resources to give an accurate decision especially in the case of breast cancer. Automated system can provide faster and more reliable method of diagnosis and prediction.

In [23], is proposed a system that is based on Artificial Neural Network that helps physicians to classify breast cancer tumors, the model were built using Matlab 2010a where is used the data set provided by the university of California. The data set consists of 32 attributes, 10 of them are real-valued features computed for each cell, resulting in 30 input features for 569 images representing 357 benign and 212 malignant cases. From this data is extracted 424 records (212 malignant, 212 benign), 360 training records, 42 testing records and 42 validation records. The work presented, first start with one hidden layer and then is incremented the number of hidden layer until finding the correct number. The input layer is 30 features the output layer had one output layer can be malignant or benign. The results show that for an input features 30, one hidden layer (with 16 features) and one output layer, the accuracy of prediction is about 97.62%. The number of hidden layers is increased to 2, 3, 4, 5, best results are obtained for three hidden layer with 97.62%. For five hidden layer, the accuracy is decreased to 96.90%. A comparison between the works presented in [20, 22, 23] is explained in Table 2.

Table 2 Comparison between the accuracy rates of ANN for the different hidden layer

10 attributes	ANN (9, 2, 2) [20]	ANN (6, 20, 1, 1) [22]	ANN (30, 15, 8, 3, 1) [23]
Cases number	699	322	424
Accuracy rate	96.42%	99.4%	97.62%

In [24] used the dataset WBCD (Wisconsin Breast Cancer Diagnosis) to classify the tumor as benign or malignant using different neural network models, each model was tested using many combinations of parameters. The first model is based on Multilayer perception neural networks (MLP), with 1 hidden layer, 9 input, 1 output, the results show that 75% of data tested were predicted correctly (benign), only 20% were predicted correctly (malignant) and 25% were predicted false positive (are benignly classified malignant) and 80% were classified true positive (malignant classified as benign) for overall accuracy rate vary from 50.21 to 54.3% depend on the number of neuron in the hidden layer. The second model is based on radial basis function neural network where the accuracy rate is decreased to 49.79% for the same parameters as previous model. The last model was based on LVQ (learning Vector Quantization) neural networks in which the accuracy rate is increased to 94.14% for the same parameters. In [28] used MLP on WBCD data to classify ultrasound images, the accuracy rate obtained is 96.10%. In [29], it is also used MLP to classify WBCD database to show the performance of statistical neural networks on breast cancer data. 222 samples of the training data belong to benign class, 120 samples belong to malignant class. The results show that 217 samples of benign data were classified correctly in the training set, and 212 were classified correctly in the testing set, and 113 samples were classified correctly as malignant in the training data, and 116 were classified correctly as malignant in the test data that means the accuracy rate is 95.74%. [33] presented a review study of artificial intelligence technique used for cancer detection and classification that includes (breast cancer, lung cancer, brain cancer) and a comparison of accuracy rate and sensitivity and specificity is presented for each type of cancer. The results presented in [25], focus on optimization method for breast cancer classification where is used 209 images extracted from 50 patients, the images were used for training, testing and validation process, after image preparation, it is used two classification techniques that are logistic regression (LR) [26], and back propagation neural networks (BPNN) [26], the results show that the number of features used in LR were much higher than that used in (BPNN), however, the classification results were better in BPNN than in LR. The accuracy obtained in BPNN exceeds 93% for 209 images (96 are benign—110 are malignant).

3.3 Fuzzy C-Means

A good amount of work use Fuzzy C-mean clustering algorithm. Fuzzy C-mean classifier is based on pattern Recognition model, as in any classifier data is divided to training set and testing set, but in fuzzy C-mean, the pattern recognition process is based on finding the similarities with known structure in data. Data are organized in clusters and to each input data are assigned a cluster that has high degree of similarities. In [30], it is used fuzzy C-mean to classify 683 samples, 444 benign and 239 malignant, the output is two classes: benign and malignant. 683 samples were divided into testing samples and training samples 400 instances were used of test data, 200 benign instances and 200 malignant instances. The results show that the accuracy rate in classification for true positive is 100%, for true negative is 87% and only 13% of data classified were false positive. Results presented in the paper indicate that the fuzzy C-mean is the best algorithm for classification given the data and parameters used, total accuracy is calculated as follow:

$$\text{Accuracy} = \frac{TP + TN}{TP + TN + FP + FN}$$

where TP = true positive that means sick, TN means true negative that means healthy, FP = false positive that means negative identified as positive (healthy identified as sick), FN = false negative that means positive identified as negative (sick identified as healthy) So:

$$Accuracy = \frac{100 + 87}{100 + 87 + 13} = 93.5\%.$$

In [31] used SVM (Support Vector Machine) algorithm to classify the data obtained from public health area in Spain, where 2% of the total population identified with breast cancer. Data are ordered in matrix n*n with dimensions (144, 169, 196), the simple size is 100 persons. Three mathematical models are used to validate the performance of SVM algorithm in each model is calculated sensitivity, specificity, and accuracy where 144 samples were used for training and 100 samples were used for testing. Sensitivity is statistical measure of the performance of binary classification test that is calculated as the ratio of actual positive cases that are correctly identified. Specificity is the actual negatives correctly identified.

$$\text{Sensitivity} = \frac{TP}{TP + TN}, \text{ specificity} = \frac{TN}{TP + TN}$$

Results show that if the specificity decreases the sensitivity increases and the accuracy decreases. While we are interested in the accuracy of prediction, we can state that the accuracy using SVM is about 80%, in comparison with fuzzy C-mean and ANN where the accuracy in fuzzy C-mean in some works is about 93.5%, and in ANN, in some works, can reach 99.4. Table 3 presents a comparison between ANN, SVM, Fuzzy C-mean and MLP accuracy in the works presented in this paper.

3.4 K-Nearest Neighbors

This algorithm is one of the machine learning algorithms that does not require a learning phase, it is based on different parameters, and the most important one is k that represents the distance that separates the neighbors to the chosen element. That means the classification method is based on nearest neighbors where is defined the distance between elements as the length between two points. Then, is classified the

Table 3 Comparison between accuracy rates of different artificial intelligence tools used for breast cancer prediction

AI technology	Accuracy rate			
Neural networks	96.42% [20]	99.4% [22]	97.62 [23]	93% [26]
SVM	80% [31]	–	–	–
Fuzzy C-mean	93.5% [30]	–	–	–
MLP	50.1–54.3 [24]	96.1% [28]	95.74% [29]	–
Decision tree	ID3 94.56 [17]	J48 92.87 [20]	–	–

similar elements at the same distance to a determined point in a single class. For example, consider x is our used point, k is the distance to that point and N is the number of similar elements at distance k from x, N form a class C, k is considered the distance between x and any point in the class C. This operation is repeated recursively for other points in the environment considered. The choice of k is depend on the data. [34] proposed a k-nn algorithm for breast cancer detection, where is considered 699 clinical cases, 458 were benign and 241 were malignant. k is defined in three methods the first is the nearest rule in which the new element will be assigned to the majority class among the nearest neighbors. The second rule is the random rule in which is considered the value of k, if k is even so, the class of new element will be that of nearest neighbors and it will be done randomly, k is calculated in different methods like Euclidean distance, cosine distance, correlation distance. The third method is the consensus in which the new element will only be affected if all nearest neighbors are of the same class. The classification accuracy is calculated for each method used rate for different value of k. The results show that the highest classification accuracy rate 98.70% is reached by the algorithm that uses the Euclidean distance with value $k = 1$, however, the accuracy rate is decreased where k is increased.

4 Conclusion and Future Works

Our work in this paper is to study the efficiency of CAD (Computer Aided Detection) in early breast cancer for that we have seen many researches that use artificial intelligence in early detection of breast cancer, and we have made comparisons of the results of the algorithms used to devise the best in terms of the degree of accuracy of classification, the results showed that ANN gives the best accuracy rate (93–99.4%) for different scenarios, Fuzzy C-mean can reach accuracy rate of 93.5%, decision tree algorithm performs in a good manner but it depends on the sample size (92.87–94.5%). SVM accuracy rate about 80%, however, MLP gives a wide range of variety in accuracy rate in the works studied (50.1–96.1%). Our future work is to decision tree algorithm as a classification method in different works like spam image detection to study if the algorithm performance will change given the input parameters.

References

1. imaginis.ccm/breast-cancer-resource-cent
2. Ganesh S, Rahul D, Jyotsana S, Piush S (2010) Various types and management cf breast cancer: an overview. J Adv Pharm Technol Res 1(2):109–126. Available at link: https://www.ncbi.nlm.nih.gov/pmc/articles/pmc3255438
3. Khatib O (2006) Guidelines for early detection and screening of breast cancer. World Health Organization, Regional for the Eastern Mediterranean. Available at: https://apps.who.int/iris/handle/10665/119805. ISBN: 978-92-9021-406-9
4. Niknejad M (2022) Breast imaging reporting and data system (BI-RADS). American college of radiology. Radiopaedia. Available at: http://www.breast-imaging-reporting-and-data-system-bi-rads
5. NgamaT (2006) World health organization cancer priorities in developing countries. J Oncol 17(8):Viii9–viii4 [pubmed]
6. Sholman L, Willett W, Sievers A, Kraul F (2010) Breast cancer in developing countries opportunities for improved survival. J Oncol. https://doi.org/10.1153/2020/595167. PM-CID. Available at www.ncbi.clm.gov/articles/PMC3021855
7. Silverman B, Culver H, Bray F et al, Survey of breast cancer incidence in the middle East and the United States. Int J Women's Health Wellness 3(1). https://doi.org/10.23937/2474-1353/1510045
8. American Cancer Society Breast Cancer Facts & figures 2017–2018. Atlanta
9. Seely J, AlHassan T (2018) Screening for breast cancer in 2018 what should we be Doing today? Curr Oncol: Can Cancer Res J 25(suppl). PMCID: PMC-6001765
10. Martin L (2017) Breast calcification, symptoms, causes, tests and treatments. WebMD Med Ref. Available at: https://www.webmd.com/women/guide/breast-calcication-symptoms-causes-treatments-2.
11. Michaelson J, Satija S, Moor R et al (2002) The pattern of breast cancer screening utilization and its consequences. Cancer a4(1):37–43. PubMed.gov. PMID: 11815958. https://doi.org/10.1002/cncr.10154
12. Baines CJ, Mc Farlane DV, Miller AB (1990) The role of the reference radiologists: estimates of interobserver agreement and potential delay in cancer detection in national screening study. Invest Radial 25(a):971–976. PubMed.gov. PMID:2211054. https://doi.org/10.1097/00004424-199009000-00002
13. Bandyopadhyay S (2010) Pre-processing of mammogram images. Int J Eng Sci Technol 2(11):6753–6758
14. Sharma M, BDubey R, KGupta S (2012) Feature extraction of mammograms. Int J Adv Comput Res 2(5):3
15. Breast Cancer Statistics World Cancer Research fund: http://www.wcrf.org/dietandcancercancer-trendsbreast-cancerstatistics. Source from: Bray F, Ferlay J et al (2018) A global cancer statistics. GloboCAN estimates of incidence and mortality worldwide for 36 cancer in 185 countries. GloBoCAN 2018 Database
16. Han J, Kamber M (2011) Data mining concepts and technique. Morgan Kauffman Publishers, 3rd edn. ISBN-10:9780123814791
17. Sumbaly R, Vishnusvi N, Jeyalatha S (2014) Diagnosis of breast cancer using decision tree data mining technique. Int J Comput Appl 98(10)
18. Bhargava N, Sharma G, Bhargava R, Mathuna M (2013) Decision tree analysis on J48 algorithm for data mining. Proc Int J Adv Res Comput Sci Softw Eng 3(6)
19. Witten IH, Frank E et al (1999) Weka: practical machine learning tools and techniques with java implementation. Hamilton, New Zeeland, University of Waikato, Department of Computer Science
20. Autsuo H (2018) Diagnosis of breast cancer using decision tree and artificial neural networks algorithms. Int J Comput Appl Technol Res 7(01). ISSN: 2319-8656
21. Suzuki K (2013) Artificial neural networks: architectures and applications, 1st edn. ISBN: 978-953-51-0935-8. Janeza Trdine 9, 51000 Rijeka-Croatia

22. Nadeem T (2017) Breast cancer detection using artificial neural networks. J Mol BiomarkS Diagn 9(1). https://doi.org/10.4172/2155-9929.100037/
23. Al-Dreabi E, Otoom M, Salah B, Hawamdeh Z (2017) Automated detection of breast cancer using artificial neural networks and fuzzy logic. Int J Sci: Basic Appl Res (IJSBAR) 35(3):109–120
24. Janhel R, Shukla A, Tiwari R, Kala R (2010) Breast cancer diagnosis using artificial neural network models. In: The 3rd international conference on information sciences and interaction sciences. Chengdu, China. https://doi.org/10.1109/ICICIS.2010. 5534716
25. Hadidi M, Hanahnah M (2016) Breast cancer detection using K-nearest neighbor machine learning algorithm. In: 9th international conference on development in e-systems engineering. https://doi.org/10.1109/DoSE,2016.8
26. Shwartz S, Ben-David S (2014) Understanding machine learning from theory to algorithms, 1st edn. Cambridge University Press. ISBN: 978-1-107-05713-5
27. Basha S, Prasad S (2009) Automatic detection of breast cancer mass in mammograms using morphological operators and fuzzy C-means clustering. J Theor Appl Inf Technol
28. Jung I-S, Thapa D, Wang G-N (2005) Neural network based algorithms for diagnosis and classification of breast cancer tumor. In: Computational intelligence and security, vol 3801. Springer, Berlin, pp 107–114
29. Swathis S, Rizwana S, Babu GA, Kumar PS, Sarma P (2012) Classification of neural networks structures for breast cancer diagnosis. Int J Comput Sci Commun 3(1):227–231
30. Muhic I (2013) Fuzzy analysis of breast cancer disease using fuzzy C-means and pattern recognition. https://doi.org/10.21533/scjournal.v2il.45
31. Menendez L, Decos Juez F, Sanchez Lasheras F, Riesgo JA (2010) Artificial neural networks applied to cancer detection in a breast screening programm. In: Mathematical and computer modelling, vol 52. Elsevier, pp 983–991
32. Breast Cancer Statistics. Approved by Cancer.Net Editorial Board in 01/2020. https://www.cancer.net/cancer-types/breast-cancer/statistics
33. Shamasneh A, Onaidellah U (2012) Artificial intelligence techniques for cancer review study. Eur Sci J. https://doi.org/10.19044/esj.2016.v13n3p342
34. Medjahed SA, Saadi T, Benyettou A (2013) Breast cancer diagnosis by using k-nearest neighbor with different distances and classification rules. Int J Comput Appl 62(1)
35. Ajithkumar TV, Hatcher HM (2010) Specialist training in oncology book. Chapter 10, 1st edn, pp 115–153. Science Direct. ISBN:9780723434580

Chapter 8
Breast Cancer Classification and Survival Prediction Using Proteomic Analysis

Kirti Aggarwal, Anuja Arora, and Joel Azzopardi

1 Introduction

Breast cancer is the top cancer diagnosed in women, after skin cancer, and it is a major contributor to cancer-related fatalities. However, some men are also diagnosed with it but it is most common in women. Early detection and accurate classification of breast cancer are critical for the effective treatment and survival of patients [1]. At present, the method used for the diagnosis of breast cancer is too invasive; a small tissue lump is taken out and examined and tested by the specialist in labs. This technique is called breast biopsy. This technique is quite accurate but cancer grows more rapidly after this test [2]. Breast cancer stage can be identified by the characteristics like shape, size of the infected tissue, tumor grade, estrogen- and progesterone-receptor status, amount of HER2 proteins generated, Oncotype dx score, and the rate at which it is spreading to lymph nodes and other distant organs [3]. The five most widely recognized molecular subtypes of breast cancer are depicted in Fig. 1.

1.1 Luminal A

Luminal ER-positive and HER2 negative. The Luminal A subtype is defined by its high expression of estrogen receptors (ER) and low expression of human epidermal growth factor receptor 2 (HER2). It has a low level of ki protein. Its tumor grade is

K. Aggarwal (✉) · A. Arora
Jaypee Institute of Information Technology, Noida, India
e-mail: aggarwalkirti25@gmail.com

J. Azzopardi
Department of AI, Faculty of ICT, University of Malta, Msida, Malta

M. Sahni et al. (eds.), *Novel Developments in Futuristic AI-based Technologies*, Algorithms for Intelligent Systems, https://doi.org/10.1007/978-981-99-3076-0_8

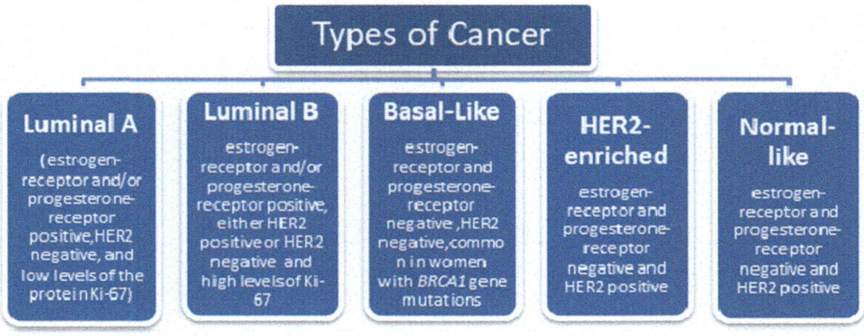

Fig. 1 Types of cancer

either 1 or 2, which is quite a low grade [4]. They grow very slowly and can very easily be predicted. These tumors tend to have a better prognosis and respond well to hormonal therapy.

1.2 *Luminal B*

This subtype, similar to Luminal A, is characterized by elevated HER2 expression and occurs when Luminal B cells begin to grow in the mammary ducts and inner lining of the breast. These are ER-positive and HER2-positive or HER2-negative [5]. The prediction of these cells is less accurate as compared to luminal A. Its tumor is larger and carries high levels of Ki-67. These tumors are often more aggressive and may require a combination of hormonal and targeted therapies.

1.3 *Triple-Negative*

Triple-negative breast cancer is defined by the absence of estrogen receptor (ER), progesterone receptor (PR), and HER2 expression, making it both ER- and PR-negative. It is also HER2 negative [6]. It has a subset named Basal-Like. These tumors tend to be more aggressive and have a poorer prognosis, with limited treatment options available. Triple-negative contributes to about 20% of total breast cancers. Triple-negative breast cancer is prevalent among women with BRCA1 gene mutations.

1.4 HER2-Enriched

This subtype is characterized by high HER2 expression and low ER expression. HER2-enriched breast cancer can be either HER2-positive or HER2-negative [7]. This can be predicted easily at younger stages but as cancer grows, it becomes difficult to diagnose it. This cancer grows faster than the other types of breast cancer. Although these tumors tend to be more aggressive and have a less favorable prognosis, they can be treated with therapies aimed at HER2.

1.5 Normal-Like

This subtype is characterized by gene expression patterns that are similar to those seen in normal breast tissue. Normal-like breast cancer is HER2-negative cancer. It contains very less Ki-67 protein [8]. These tumors tend to have a favorable prognosis, but they may still require treatment. Diagnosis of these cells is easy as compared to luminal B but difficult as compared to luminal A.

Proteomic analysis, which is the study of the protein content and function in a cell, tissue, or organism, offers a promising approach to the classification and survival prediction of breast cancer [9]. By analyzing proteins, proteomic analysis can reveal valuable insights into the biological processes and changes occurring in breast cancer cells, facilitating the identification of biomarkers and the creation of new diagnostic tools. The identification of specific protein patterns and markers can help to classify different subtypes of breast cancer, which can have significant implications for personalized treatment and prognosis [10].

In addition to classification, proteomic analysis can also be used for survival prediction in breast cancer patients [11, 12]. By analyzing the protein content and function in patient samples, it is possible to identify prognostic markers that are associated with patient outcomes, such as disease progression and recurrence [13]. These markers can be used to develop prognostic models that can accurately predict the survival of the patient and help to guide treatment decisions.

Additionally, proteomic analysis can be used to monitor treatment response and predict treatment outcomes. For example, studies have shown that changes in the protein profile of a patient's tumor can indicate resistance to treatment and the need for alternative therapeutic strategies. Therefore, proteomic analysis is a valuable asset for the diagnosis, treatment, and management of breast cancer [14]. Yet, further research is needed to fully understand the utility of proteomic analysis in the clinical setting.

2 Literature Study

Breast cancer classification and survival prediction using proteomics analysis is a growing area of research. The proteomic analysis involves the study of the protein composition of a sample and can provide insights into the molecular mechanisms involved in cancer development and progression [15]. Proteomic analysis can provide insights into the molecular mechanisms that drive cancer development and progression [16]. Scientists can uncover proteins that are specifically altered in cancer and could be targeted for therapy by examining the differences in protein expression between normal and malignant tissue through proteomic analysis [17]. For instance, an increased expression of growth factor receptors like epidermal growth factor receptor (EGFR) has been associated with a more virulent form of breast cancer.

Breast cancer is a multifaceted illness, and the capacity to categorize it into various subtypes based on molecular features is critical for delivering individualized treatment approaches [18, 19]. The proteomic analysis provides a formidable means for classification as it enables researchers to examine the protein expression patterns in tumor tissue and differentiate between the various subtypes of breast cancer [20]. For example, proteomic analysis has been employed to categorize breast cancer into luminal A and B subtypes, HER2-positive, and triple-negative subtypes based on variations in protein expression. This information is valuable for guiding treatment decisions, as each subtype of breast cancer has a different biological behavior and response to therapy [21].

In addition to classification, proteomic analysis can also be used to predict patient survival outcomes [22]. Researchers have identified specific protein markers that can serve as indicators of a poor prognosis and help predict patient survival. For example, an elevated expression of the tumor protein 53 (p53) has been linked to a less favorable outcome in breast cancer [23]. By monitoring changes in the protein expression profile of a patient's tumor over time, proteomic analysis can also be used to monitor treatment response and predict treatment outcomes. This information is valuable for guiding treatment decisions and improving patient outcomes.

Another important use of proteomic analysis in breast cancer is to monitor treatment response and predict treatment outcomes [24]. For example, changes in the protein expression profile of a patient's tumor can indicate whether a particular treatment is working, and if not, help guide the selection of more effective therapy. This information can also be used to personalize treatment plans for individual patients based on their unique tumor biology [25].

The prediction of breast cancer using proteomic analysis can also be done using machine learning [26] and data mining techniques [27]. Some of the research work done in this field are as follows:

Asri et al. in their research work stated that data mining algorithms can be utilized for early cancer prediction, allowing for necessary measures to be taken to prevent it. Early detection is crucial, as cancer is more readily treatable in its initial stages [28]. In this paper, Asri et al. have compared various models and ranked them based on their performance and the accuracy of the prediction. The dataset they worked on was from Wisconsin Breast Cancer and various models like SVM, KNN, and Naive Bayes were used on this dataset. They have also used sensitivity, recall, and precision as a means to measure the performance of various models. All the classification in the paper was done using WEKA and the results indicated that Support Vector Machines (SVM) achieved the highest accuracy rate of 97.13%.

In their study, Vikas Chaurasia and Saurabh Pal compared the performance of various models such as Classification and Regression Trees (CART), neural networks, and decision trees and then evaluated the results to identify the model with the greatest accuracy for predicting breast cancer [27]. They obtained the highest accuracy of 96.84% using the SVM-RBF kernel. The dataset they used was from Wisconsin Breast Cancer.

Adam and Omar [29] worked on developing a technology-based breast cancer prediction by using BPNN (Back Propagation Neural Network algorithm). He was able to classify if the cancer was malignant or benign [29]. He performed his testing twice, once on the preprocessed data, and the second on unaltered data, from which he concluded that a higher accuracy and efficiency are obtained by the original data.

3 Proposed Idea

Information and Communication Technologies (ICT) have the potential to greatly impact cancer care. With advancements in Information Technology, the ability to manage and store vast amounts of data, known as big data, has been made possible. Data mining makes extensive use of big data applications, it has made significant changes in analyzing data and predicting it. Applications of data mining expand to its use in the health sector as it has helped in improving results of tests, predetermining whether a patient has that particular disease or not, or if it's likely to be affected by it soon. This trend has helped save a lot of people's lives.

In case of any cancer, it needs to get cured [30]. Therefore, there is an acute need for developing tools that make diagnosing cancer cheap and accurate in the earlier stage. Our goal is to create an application that addresses this issue, while also investigating the practical application of various machine learning algorithms [31], such as Random Forest, Support Vector Machine (SVM), Classification and Regression Tree (CART), and K-Nearest Neighbors (KNN), in diagnosing breast cancer through proteomic data analysis and prediction of cancer type [32]. Besides we have also predicted the probability of survival using the above-mentioned machine learning algorithms. Further, we have ensembled the algorithms by stacking them to improve the accuracy of the prediction.

4 Dataset

In this study, two datasets were used and both of these datasets were published on Kaggle [33]. The first dataset, named "77_cancer_proteomes_CPTAC_itraq.csv," is composed of proteome profiling data obtained through iTRAQ technology for 77 breast cancer samples. The data were generated by the Clinical Proteomic Tumor Analysis Consortium, a joint effort between the National Cancer Institute and the National Institutes of Health [34]. iTRAQ is a method used to label and quantify proteins in complex mixtures, and the CPTAC program aims to systematically study the proteomes of cancer cells to understand the underlying molecular mechanisms of the disease. It contains values for 12,000 proteins found in the human body that are relevant for the diagnosis of breast cancer [35]. The details included in the dataset are:

1. Sample information: information about the cancer samples analyzed, such as patient ID, tissue type, stage, and grade.
2. Protein identification information: information about the proteins that were identified, such as protein accession numbers, gene names, and sequence information.
3. Quantification data: the relative protein expression levels for each sample, as determined by iTRAQ labeling and mass spectrometry analysis.
4. Statistical analysis results: information about the statistical significance of the protein expression differences between samples, such as fold changes and p-values.

The second dataset (4.2 Clinical_data_breast_cancer.csv) consists of clinical information on breast cancer patients, including details about demographics, past medical history, physical examination outcomes, and laboratory test results. The dataset contains data such as patient ID, age, gender, tumor size, stage, grade, hormone receptor status, HER2 status, treatment history, and survival outcome. These data can be used to analyze and understand the relationship between clinical characteristics and patient outcomes in breast cancer.

5 Designing

Our project comprises two main parts. The first part includes predicting and classifying the type of breast cancer that a patient is suffering from while the second part is about predicting the vital status of the patient that is if the patient will survive or not.

5.1 Data Preprocessing

Data preprocessing is a crucial part of any project. Besides data transformation, we have also preprocessed and cleaned the data to remove the null and missing values and filter the duplicate records. We first transposed the Proteomes Dataset to make our observations in rows instead of columns. We have then changed some column names and merged both the datasets (77_cancer_proteomes_CPTAC_itraq.csv and clinical_data_breast_cancer.csv) by using inner join based on IDs. The breast cancer subtype is diagnosed using the PAM50 genomic test. This data set includes a list of proteins that are associated with the PAM50 genes. The approach we are using for variable selection is repeated lasso regression. Lasso regression shrinks the data set and generates a sparse set of predictor variables, but there is a stochastic element to the reduction and results are not always stable. To overcome this, we repeated the lasso regression 100 times and ranked the variables in order of the number of times they were selected in the final model [36].

5.2 Predicting and Classifying Breast Cancer

The objective was achieved by using four machine learning algorithms, selected based on their rankings in the research community [13, 14].

5.3 SVM

A support vector machine (SVM) is a machine learning algorithm that can perform both classification and regression analysis. It operates by finding a boundary line or hyperplane that separates the data points into different categories or predicts a numerical value based on the input data. The line or hyperplane that is found is called the maximum margin classifier, as it maximizes the margin (or distance) between the classes.

In regression analysis, the aim of SVM is to locate the line or hyperplane that best fits the given data points and at the same time maximize the gap between the data points and the line/hyperplane. The data points closest to the maximum margin classifier are known as "support vectors", and they play a crucial role in determining the position of the classifier. These support vectors are what give the classifier its "support".

SVMs are widely used in many areas, such as text classification, image classification, and bioinformatics. They are particularly useful for dealing with high-dimensional data and data with non-linear relationships between the variables.

The support vector machine (SVM) is a type of algorithm utilized in the fields of statistics and computer science for data analysis and pattern recognition. It is part of a broader group of supervised learning methods [4]. Typically, the SVM takes in a set of data and predicts the output for each input, resulting in either of two possible classes. Thus, it functions as a non-probabilistic binary linear classifier [37].

Algorithm

1. The original dataset was split into two sections: 70% was designated as training data, and the remaining 30% was set aside as testing data.
2. The SVM model was then trained on the basis of the training dataset.
3. The SVM model obtains a boundary for each class. This boundary is generally referred to as a line of the class.
4. Instances of testing data were then predicted using SVM trained model. The predicted class is the class whose line has a minimum distance from the testing data.
5. A confusion matrix was created, and the accuracy was determined by comparing the actual class to the predicted class.

5.4 CART

The CART algorithm is a decision tree model used in machine learning for both classification and regression tasks. It was developed by Breiman [38] and is commonly known as Classification and Regression Trees (CART). The algorithm works by dividing the data into subsets and making predictions based on the most relevant variables in each subset. The result is a tree-like structure that represents the different decision paths and outcomes of the analysis.

In classification problems, the goal is to predict a categorical outcome based on a set of input features. CART algorithms construct a decision tree model, showing the potential outcomes of different choices. Each internal node in the tree represents a test on a single feature, and each leaf node represents a class prediction. The prediction for a new data point is made by following the path through the tree based on the results of the feature tests.

In regression problems, the goal is to predict a continuous outcome based on a set of input features. CART algorithms build a tree-like model in the same way as for classification problems, with the difference that the leaf nodes represent continuous values instead of class predictions.

CART algorithms are widely used due to their ability to handle both categorical and numerical features, as well as their ability to provide interpretable and easily visualizable models. The CART algorithm is a type of binary decision tree that is formed by repeatedly dividing a node into two child nodes, starting with the root node that encompasses the entire training sample. It can be utilized for both classification and regression tasks. The algorithm employs the Gini index, a novel measure, to identify decision points in classification problems.

Algorithm

1. The initial dataset was divided into two parts—Training Data—70% of the original dataset, Testing Data—30% of the original dataset.
2. The CART model was then trained based on training data.
3. In CART, during the training of the model, a decision tree is built on the basis of greedy splitting.
4. Instances of testing data were then predicted on the basis of the CART-trained model. The predicted class is obtained on the basis of the decision tree building.
5. The confusion matrix was generated and the accuracy was calculated on the basis of the expected class and predicted class.

5.5 *KNN*

KNN stands for k-Nearest Neighbors. It is a type of machine learning algorithm used for both classification and regression problems. In the KNN algorithm, the idea is to predict the target variable (class label in classification or target value in regression) of a new data point based on the values of its k-nearest neighbors in the feature space. The nearest neighbors are determined based on a distance metric, such as Euclidean distance or Manhattan distance. The most common way to make predictions is to take the majority vote of the k nearest neighbors for classification problems or to average the values of the k nearest neighbors for regression problems.

KNN is a straightforward algorithm that is simple to put into practice and can be used with both small and large datasets. It is non-parametric and falls under the category of lazy learning [39]. Non-parametric means that it does not make any assumptions about the underlying data distribution and instead, the structure of the model is determined directly from the dataset. A lazy algorithm refers to a method in which model generation does not require any training data points. Instead, all the training data are utilized during the testing phase. This leads to faster model training, but slower and more resource-intensive testing. In certain cases, KNN may have longer computational time due to the need to scan all data points, and this may also require a larger memory to store the training data.

Algorithm

1. The original dataset was separated into two portions: 70% was designated as training data, and the remaining 30% was assigned as testing data.
2. The KNN model is then trained based on the training dataset.
3. KNN model classifies the test case based on K-Nearest Neighbor's classes. We have taken $K = 5$. The proximity is checked through Euclidean Distance.
4. Instances of testing data were then predicted based on the KNN model.
5. A confusion matrix was produced, and the accuracy was determined by comparing the expected class to the predicted class.

5.6 Random Forest

Random Forest is an ensemble machine learning algorithm that is used for classification and regression problems [40, 41]. It combines multiple decision trees to produce a single prediction. Random Forest involves training a large number of decision trees on bootstrapped samples of the training data. The bootstrapping process involves randomly selecting samples with replacements from the training data. Each decision tree makes its own prediction based on the features in the bootstrapped sample. In Random Forest, the final prediction is obtained by either taking the average of predictions made by each tree in the forest or by determining the most frequently predicted class in case of a classification problem (i.e. majority voting).

The utilization of multiple decision trees in Random Forest helps prevent overfitting by aggregating the predictions of individual trees to produce a stronger final prediction. Random Forest is also able to handle both categorical and numerical features and can deal with non-linear relationships between features. Random Forest is a popular and widely used algorithm, due to its ability to handle complex data structures and its good performance on a variety of problems. However, it can be computationally expensive and may not perform well when dealing with high-dimensional data.

Algorithm

1. The initial dataset was divided into two parts—Training Data—70% of the original dataset, Testing Data—30% of the original dataset.
2. The RF model was then trained on the basis of training data.
3. In RF, during the training of the model, several decision trees are built.
4. Instances of testing data were then predicted on the basis of the RF-trained model. The predicted class is the class obtained as the mean output of all the decision trees built.
5. The confusion matrix was generated and the accuracy was determined on the basis of the expected class and predicted class.

Table 1 lists the accuracy obtained using the above-described models on the dataset described in Sect. 4.

Table 1 Accuracy of algorithms

Name of the algorithm	Accuracy obtained (%)
SVM	84.7
CART	57.1
KNN	90.5
Random Forest	85.7

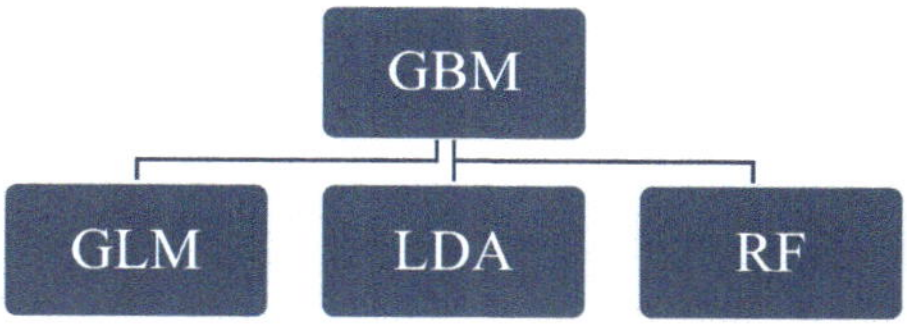

Fig. 2 Stacking using GBM as a meta-model

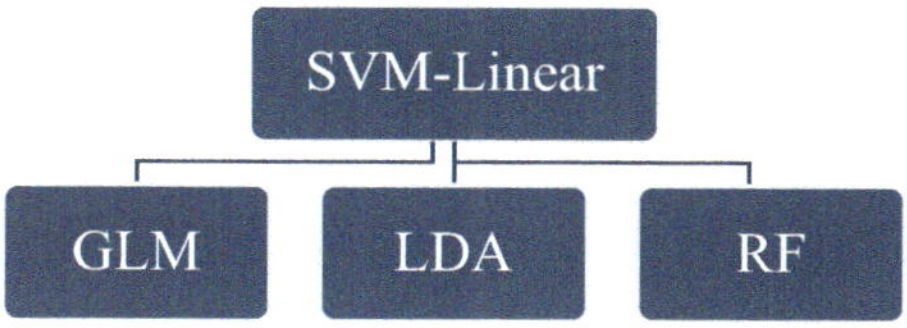

Fig. 3 Stacking using SVM-linear as a meta-model

5.7 Predicting Vital Status

Predicting Vital Status means whether the patient will live or die. We have used six algorithms to achieve our objective. Instead of using a single algorithm to predict the Vital Status, we have ensembled the algorithms by stacking them in layers to improve the accuracy. We have implemented Ensembling through stacking.

Stacking combines the predictions from several classification and regression models by using a meta-classifier or meta-regressor. The base-level models are trained on the full training dataset, and then the meta-model is trained on the output from these base models, which are treated as input features.

Linear Discriminant Analysis (LDA), Random Forest (RF), and Generalized Linear Model (GLM) have been used as base-level models. Therefore, we first trained LDA, RF, and GLM individually on the basis of our original dataset, and later we first used Generalized Boosted Regression Model (GBM) as a meta-model on the top layer that is it was trained on the output obtained by the above models. We have repeated this process with SVM-linear as a meta-model and obtained a higher accuracy percentage (Figs. 2 and 3).

5.8 Generalized Linear Model (GLM)

This model is a modified version of the linear regression model. This model works on linear predictors, generally referred to as response variables. It consists of a link and variance function. The former describes the relationship between the mean and the response variable whereas the latter tells us about the dependency of variance on the predictors. One advantage of this model over other regression algorithms is that we do not have to change the predictor variable to yield a normal distribution.

5.9 Linear Discriminant Analysis (LDA)

LDA is a generalization of Fisher's linear discriminant, which is a statistical method. In machine learning/pattern recognition, firstly a linear combo of attributes that distinguishes the classes of objects is done. This combo is further used to reduce the dimensions for linear classification.

5.10 Generalized Boosted Regression Model (GBM)

Gradient boosting machine is a very successful algorithm that is used for classification and regression problems, it builds a model which forms an ensemble of many shallow successive trees that improve and learn on their own using data from previous trees. It is useful because it constructs the model in a step-by-step manner like other boosting methods and also generalizes them by enabling the optimization of any differentiable loss function.

Algorithm

1. The initial dataset was divided into two parts—Training Data—70% of the original dataset, Testing Data—30% of the original dataset
2. LDA, GLM, and random forest were applied to obtain their predictors.
3. The predictors obtained were then used as input for the over-lying model.
4. Initially, GBM was applied as a top-layer model.
5. The above step was repeated with SVM-Linear to check whether a higher accuracy could be obtained.

The accuracy obtained by using the above-mentioned algorithm and Ensembling the two models by stacking is listed in Table 2.

Table 2 Accuracy of algorithms and stacking

Name of the algorithm	Accuracy obtained (%)
GLM	84.21
LDA	84.21
Random Forest	89.47
Stacking (Meta-Model: GBM)	89.47
Stacking (Meta-Model: SVM Linear)	94.74

6 Testing

The initial dataset was divided into two parts—Training Data—70% of the original dataset, and Testing Data—30% of the original dataset. After training the models, they were tested and the confusion matrix and accuracy were obtained. The accuracy of individual models used for predicting the vital status of patients is given in Figs. 4, 5, 6, 7 and 8. The highest accuracy that we could obtain for both predictions is listed in Table 3.

Fig. 4 Accuracy of GLM

```
Confusion Matrix and Statistics

          Reference
Prediction DECEASED LIVING
  DECEASED        0      1
  LIVING          2     16

               Accuracy : 0.8421
```

Fig. 5 Accuracy of LDA

```
Confusion Matrix and Statistics

          Reference
Prediction DECEASED LIVING
  DECEASED        0      1
  LIVING          0     18

               Accuracy : 0.9474
```

Fig. 6 Accuracy of random forest

```
Confusion Matrix and Statistics

          Reference
Prediction DECEASED LIVING
  DECEASED        0      1
  LIVING          0     18

               Accuracy : 0.9474
```

Fig. 7 Accuracy of GBM stacked with the base models

```
Confusion Matrix and Statistics

          Reference
Prediction DECEASED LIVING
  DECEASED        0      1
  LIVING          1     17

               Accuracy : 0.8947
```

Fig. 8 Accuracy of SVM Linear stacked with the base models

```
Confusion Matrix and Statistics

          Reference
Prediction DECEASED LIVING
  DECEASED        0      1
  LIVING          0     18

               Accuracy : 0.9474
```

Table 3 Testing results

	Name of algorithm	Accuracy obtained (%)
Prediction and classification of breast cancer	K-Nearest neighbors (KNN)	90.5
Prediction of vital status of a breast cancer patient	Stacking (Meta-Model: SVM)	94.74

7 Conclusion

To analyze proteomic data, we have used various data mining and machine learning algorithms. An important challenge in achieving our objective was to find an accurate and computationally efficient classifier. In this process, we have used four main algorithms: SVM, CART, KNN, and Random Forest on the Proteomic Breast Cancer dataset. We compared the efficiency of algorithms in terms of accuracy to decide the best classification model. To conclude, KNN has proven its efficiency in Breast Cancer classification with an accuracy of 90.5%. Besides in the prediction of vital status, we ensemble the algorithms to achieve the best results. The highest accuracy of prediction, i.e. 94.74% was obtained by stacking SVM as meta-model and GLM, LDA, and Random Forest as base models.

References

1. Hanash S, Taguchi A (2011) Application of proteomics to cancer early detection. Cancer J 17(6):423–428. NIH Public Access. https://doi.org/10.1097/PPO.0b013e3182383cab
2. Łukasiewicz S, Czeczelewski M, Forma A, Baj J, Sitarz R, Stanisławek A (2021) Breast cancer—epidemiology, risk factors, classification, prognostic markers, and current treatment strategies—an updated review. Cancers 13(17):4287. Multidisciplinary Digital Publishing Institute. https://doi.org/10.3390/cancers13174287
3. Kalita-de Croft P, Straube J, Lim M, Al-Ejeh F, Lakhani SR, Saunus JM (2019) Proteomic analysis of the breast cancer brain metastasis microenvironment. Int J Mol Sci 20(10):2524. https://doi.org/10.3390/ijms20102524
4. Ignatiadis M, Sotiriou C (2013) Luminal breast cancer: from biology to treatment. Nat Rev Clin Oncol 10(9):494–506. https://doi.org/10.1038/nrclinonc.2013.124

5. Creighton CJ (2012) The molecular profile of luminal B breast cancer. Biologics 6:289–297. https://doi.org/10.2147/BTT.S29923
6. Rasskazova EA (2014) Triple-negative breast cancer. PA Herzen J Oncol 3(5):65–70. https://doi.org/10.1056/NEJMRA1001389
7. Schettini F et al (2020) HER2-enriched subtype and pathological complete response in HER2-positive breast cancer: a systematic review and meta-analysis. Cancer Treat Rev 84. https://doi.org/10.1016/j.ctrv.2020.101965
8. Bu W et al (2011) Keratin 6a marks mammary bipotential progenitor cells that can give rise to a unique tumor model resembling human normal-like breast cancer. Oncogene 30(43):4399–4409. https://doi.org/10.1038/onc.2011.147
9. Qin XJ, Ling BX (2012) Proteomic studies in breast cancer (Review). Oncol Lett 3(4):735–743. Spandidos Publications. https://doi.org/10.3892/ol.2012.573
10. Neagu A-N et al (2022) Proteomics-based identification of dysregulated proteins in breast cancer. Proteomes 10(4):35. https://doi.org/10.3390/proteomes10040035
11. Neagu A-N et al (2021) Proteomics and its applications in breast cancer. Am J Cancer Res 11(9):4006–4049. Accessed 04 Feb 2023 [Online]. Available: /pmc/articles/PMC8493401/
12. Ősz Á, Lánczky A, Győrffy B (2021) Survival analysis in breast cancer using proteomic data from four independent datasets. Sci Rep 11(1). https://doi.org/10.1038/s41598-021-96340-5
13. Baak JPA, Path FRC, Hermsen MAJA, Meijer G, Schmidt J, Janssen EAM (2003)Genomics and proteomics in cancer. Eur J Cancer 39(9):1199–1215. Pergamon. https://doi.org/10.1016/S0959-8049(03)00265-X
14. Tan HT, Lee YH, Chung MCM (2012) Cancer proteomics. Mass Spectrom Rev 31(5):583–605. https://doi.org/10.1002/mas.20356
15. Abramovitz M, Leyland-Jones B (2006) A systems approach to clinical oncology: Focus on breast cancer. Proteome Sci 4(1):1–15. https://doi.org/10.1186/1477-5956-4-5
16. Qin XJ, Ling BX (2012) Proteomic Stud Breast Cancer (Review) 3(4):735–743. https://doi.org/10.3892/ol.2012.573
17. Neagu AN et al (2021) Proteomics and its applications in breast cancer. Am J Cancer Res 11(9):4006–4049. Accessed 04 Feb 2023 [Online]. Available: http://www.ncbi.nlm.nih.gov/pubmed/34659875%0A, http://www.pubmedcentral.nih.gov/articlerender.fcgi?artid=PMC8493401
18. Yanovich G, Agmon H, Harel M, Sonnenblick A, Peretz T, Geiger T (2018) Clinical proteomics of breast cancer reveals a novel layer of breast cancer classification. Cancer Res 78(20):6001–6010. https://doi.org/10.1158/0008-5472.CAN-18-1079
19. Lam SW, Jimenez CR, Boven E (2014) Breast cancer classification by proteomic technologies: Current state of knowledge. Cancer Treat Rev 40(1):129–138. W.B. Saunders. https://doi.org/10.1016/j.ctrv.2013.06.006
20. Tyanova S, Albrechtsen R, Kronqvist P, Cox J, Mann M, Geiger T (2016) Proteomic maps of breast cancer subtypes. Nat Commun 7(1):1–11. https://doi.org/10.1038/ncomms10259
21 An R et al (2022) Integrative analysis of plasma metabolomics and proteomics reveals the metabolic landscape of breast cancer. Cancer Metab 10(1).https://doi.org/10.1186/s40170-022-00289-6
22. Asleh K et al (2022) Proteomic analysis of archival breast cancer clinical specimens identifies biological subtypes with distinct survival outcomes. Nat Commun 13(1):1–19. https://doi.org/10.1038/s41467-022-28524-0
23. Zeidan BA, Townsend PA, Garbis SD, Copson E, Cutress RI (2015) Clinical proteomics and breast cancer. Surgeon 13(5): 271–278. Elsevier. https://doi.org/10.1016/j.surge.2014.12.003
24. Latterich M, Abramovitz M, Leyland-Jones B (2008) Proteomics: New technologies and clinical applications. Eur J Cancer 44(18):2737–2741. https://doi.org/10.1016/j.ejca.2008.09.007
25. He et al (2020) A review on methods for diagnosis of breast cancer cells and tissues. Cell Prolif 53(7). Blackwell Publishing Ltd.https://doi.org/10.1111/cpr.12822
26. Gayathri BM, Sumathi CP, Santhanam T (2013) Breast cancer diagnosis using machine learning algorithms–a survey. Int J Distrib Parallel Systs 4(3):105–112. https://doi.org/10.5121/ijdps.2013.4309

27. Chaurasia V, Pal S, Tiwari BB (2018) Prediction of benign and malignant breast cancer using data mining techniques. J Algorithm Comput Technol 12(2):119–126. https://doi.org/10.1177/1748301818756225
28. Asri H, Mousannif H, Al Moatassime H, Noel T (2016) Using machine learning algorithms for breast cancer risk prediction and diagnosis. Procedia Comput Sci 83:1064–1069.https://doi.org/10.1016/j.procs.2016.04.224
29. Adam A, Omar K (2006) Computerized breast cancer diagnosis with Genetic Algorithm and Neural Network. In: Proceeding of the 3rd international conference on artificial intelligence and engineering technology (ICAIET), pp 22–24. Accessed 04 Feb 2023. [Online]. Available: https://citeseerx.ist.psu.edu/document?repid=rep1&type=pdf&doi=fbf8705bdc07c0a09f1d947f0ccb6ebdb55a855c
30. Almarzooq R, Alrayes A, Alaradi H, Abdulla H (2018) Molecular subtypes of breast cancer. Bahrain Med Bull 40(4):222–225. https://doi.org/10.36106/ijsr/1634010
31. Taguchi T, Sugiura M, Hamada Y, Miwa I (1998) In vivo formation of a Schiff base of aminoguanidine with pyridoxal phosphate. Biochem Pharmacol 55(10):1667–1671. https://doi.org/10.1016/S0006-2952(98)00010-0
32. Tahmooresi M, Afshar A, Bashari Rad B, Nowshath KB, Bamiah MA (2018) Early detection of breast cancer using machine learning techniques. J Telecommun, Electron Comput Eng 10(3–2):21–27. Accessed 04 Feb 2023. [Online]. Available: https://jtec.utem.edu.my/jtec/article/view/4706
33. "Breast Cancer Proteomes | Kaggle." https://www.kaggle.com/datasets/piotrgrabo/breastcancerproteomes. Accessed 10 Nov 2021
34. "Breast Cancer Proteomes | Kaggle." https://www.kaggle.com/piotrgrabo/breastcancerproteomes. Accessed 10 Nov 2021
35. Mertins P et al (2016) Proteogenomics connects somatic mutations to signalling in breast cancer. Nature 534(7605):55–62. https://doi.org/10.1038/nature18003
36. Wu X et al (2008) Top 10 algorithms in data mining. Knowl Inf Syst 14(1):1–37. https://doi.org/10.1007/s10115-007-0114-2
37. Qinli Z, Shitong W, Qi G (2007) A novel SVM and its application to breast cancer diagnosis. In: 2007 1st international conference on bioinformatics and biomedical engineering, ICBBE, pp 633–636. https://doi.org/10.1109/ICBBE.2007.165
38. Breiman L, Friedman JH, Olshen RA, Stone CJ (2017) Classification and regression trees. Classification Regres Trees 45:1–358. https://doi.org/10.1201/9781315139470
39. Noble WS (2006) What is a support vector machine? Nat Biotechnol 24(12): 1565–1567. Nature Publishing Group. https://doi.org/10.1038/nbt1206-1565
40. Pavlov YL (2019) Random forests, vol 45, no 1. Springer, Berlin. https://doi.org/10.4324/9781003109396-5
41. Breiman L (2001) Random forests. Mach Learn 45(1):5–32. https://doi.org/10.1023/A:1010933404324

Chapter 9
An Improved Approach in Cloud-Assisted IoT Infrastructure with Anonymous Key Issuing

Nishant Doshi and Chintan Patel

1 Introduction

Upon the evolvement of technology, privacy and security become the prime concern for any real-time applications. Through cryptography, one can achieve the security to withstand attacks. IoT is a key technology of the twenty-first century due to its pluses like hostile environment, low cost, low maintenance, etc. [1–5]. Various solutions were proposed for security concerns in IoT, as depicted in the literature [6–12]. Further, it will lead to the cloud-assisted IoT system by merging the power of cloud-based architecture [13, 14]. Figure 1 shows the difference between conventional and encrypted cloud-assisted IoT systems.

One of the primitive solutions to provide effective multicast support in encryption is to use the ABE [15]. CP-ABE [16–18], a type of ABE, becomes the choice of researchers as it provides flexibility to the sender for selecting the intended recipients. Afterwards, many approaches [16–41] were mentioned to improve the CP-ABE scheme. Recently, in [42], the authors have proposed a scheme for cloud-assisted IoT systems having constant length ciphertext. It is effective compared to variable length ciphertext due to the avoidance of communication and computation overhead on the system's end users. Thus, in this research, we are focusing on the constant-length approach. AKI (Anonymous Key issuing) is one of the primitive features of the CP-ABE scheme as it helps to protect against honest-cum-curious central authority to know the identity of the users.

N. Doshi
Department of Computer Science, Pandit Deendayal Energy University, Gandhinagar, India

C. Patel (✉)
Department of Computer Science, University of Sheffield, Sheffield, UK
e-mail: Chintan.p592@gmail.com

M. Sahni et al. (eds.), *Novel Developments in Futuristic AI-based Technologies*, Algorithms for Intelligent Systems, https://doi.org/10.1007/978-981-99-3076-0_9

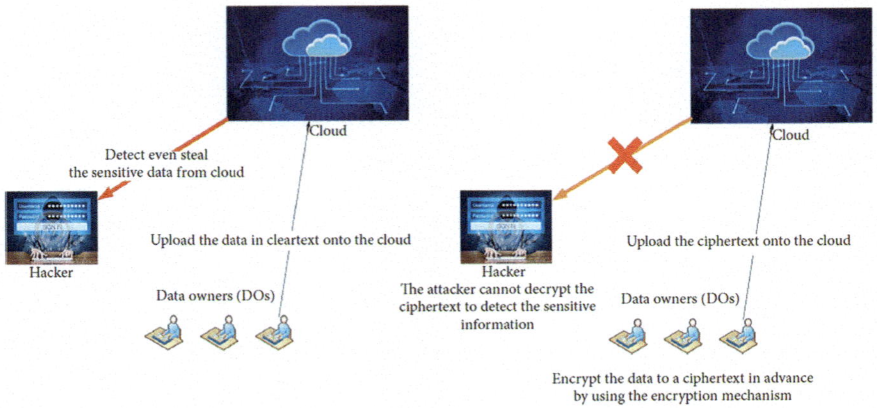

Fig. 1 Cloud-assisted IoT infrastructure [42]

This also helps protect against key collusion and key escrow attacks. The scheme of [42] fails to provide this feature, and this research claimed an efficient method to overcome the limitations of the earlier schemes.

1.1 Our Contribution

As evident from the literature survey, there is a need for a single system having all-in-one features from AKI, constant length ciphertext. The proposed scheme works in the single authority environment where all trust is kept upon a single trusted server. The proposed scheme required policy attributes in ciphertext to be a subset of the user's secret key attributes for successful decryption.

1.2 Paper Organization

Section 2 deals with the basic terminology used in the remaining paper. Section 3 gives the mentioned scheme with a details analysis of each algorithm. Section 4 gives a comparative analysis of the proposed against earlier schemes. The conclusion is given at the end.

2 Preliminaries

Definition 1 (*Bilinear map*). Consider cyclic group G_1, G_2 of multiplicative generators g_1, g_2, and g_3. It is function $e : G_1 \times G_2 \rightarrow G_3$ having the following requirements, Bi-linearity, Non-degeneracy, and Efficiency are used in the proposed scheme.

Definition 2 (*DLP*) With g and h, gets $x \in Z_p$ such that $h = g^x$ if exist else random value.

3 The Proposed Scheme

The scheme requires the following methods.

Set-up: It executes by CA as follows. Takes G_0 of prime order p with generator δ. Gets $y, \beta \in_R Z_p$ Compute $h = \delta^\beta, Y = e(\delta, \delta)^y MPK = G_0, \delta, h, Y, MSK = (\beta, y)$

- For each system attribute i with value j.
- Takes $\alpha_{i,j} \in_R Z_p$.
- Gets $PK_i = \delta^{\sum v_{i,j}\alpha_i}$ and $SK_i = \sum v_{i,j}\alpha_i$

KeyGen (MSK, u): It run by CA to compute secret key of user. Selects $r \in_R Z_p$. Gets $SK_u = \delta^{\sum v_{i,j\in L_u}} \alpha_i(y+r)/\beta$, $PK_u = \delta^r$ with attribute list $L_u = \phi$

KeyGenAttributes ($\mathbf{PK_u}$, u, $\mathbf{L_u}$): This algorithm run by CA.

- (before) User generates $x \in_R Z_p$ and give $PK' = (\delta^r)^{\prod v_{i,j\in L_u}\alpha_i x}$ to AA.
- (in) AA will give $D_i = (\delta^{rx})^{\prod v_{i,j\in L_u}} \alpha_i$ and $L_u = L_u + i$ to user.
- (after) User compute $(D_i)^{1/x} = \delta^{r \prod_{v_{i,j\in L_u}} \alpha_i}$.

As the user is giving the PK', this will protect the anonymity of user and maintain the AKI property.

Encrypt (M, W, $\mathbf{PK_1}$, $\mathbf{PK_2}$,…, $\mathbf{PK_N}$): Sender runs the following steps. Selects$s \in_R Z_p$. Gets$C_1 = MY^s$,$C_2 = \delta^s$,$C_3 = \left(\prod_{t\in W} \delta^{\prod_{v_{i,j\in L_u}} \alpha_i}\right)^s$, $C_4 = (h)^s = \delta^{\beta s}$

- Final ciphertext $CT = \{C_1, C_2, C_3, C_4, W\}$

Decrypt (SK, CT): The receiver runs the following steps to get message M.

$$= \frac{C_1 \cdot e(\delta^r, C_2) \cdot e(C_3, \delta^r)}{e\left(C_4, \delta^{\frac{y+r}{\beta}}\right) \cdot e\left(C_2, \left(\prod_{t\in AS} \delta^{\prod_{v_{i,j\in L_u}} \alpha_t}\right)^r\right)}$$

$$= M$$

4 Analysis

In Table 1, we have given the comparative analysis based on the size of effective parameters with the nature of the policy. In Table 2, we have extended the comparisons to the computation time with earlier schemes. In Table 3, we have done a comparative analysis based on various important features. As per the conclusion from all tables, the proposed scheme outperforms as compared to earlier schemes.

Table 1 Size of ABE schemes

Scheme	MPK	MSK	SK	CT
[16]	–	–	$O(\mathfrak{R})|\mathfrak{g}|$	$O(\mathcal{R}_1)|\mathfrak{g}|$
[17]	–	–	$O(\mathfrak{R})|\mathfrak{g}|$	$O(\mathcal{R}_1)|\mathfrak{g}|$
	–	–	$O(\mathfrak{R})|\mathfrak{g}|$	$O(\mathcal{R}_1)|\mathfrak{g}|$
[41]	$O(\mathfrak{R})|\mathfrak{g}|$	$O(n)|Z|$	$O(\mathcal{R}_2)|\mathfrak{g}|$	$O(1)|\mathfrak{g}|$
[19]	$O(1)|\mathfrak{g}|$	$O(1)|\mathfrak{g}|$	$O(\mathcal{R}_2)|\mathfrak{g}|$	$O(1)|\mathfrak{g}|$
[31]	$O(\mathfrak{R})|\mathfrak{g}|$	$O(\mathfrak{R})|Z|$	$O(\mathcal{R}_2)|\mathfrak{g}|$	$O(1)|\mathfrak{g}|$
[42]	$O(1)|\mathfrak{g}|$	–	$O(\mathcal{R}_2)|\mathfrak{g}|$	$O(1)|\mathfrak{g}|$
Our work	$O(\mathfrak{R})|\mathfrak{g}|$	$O(1)|Z|$	$O(\mathcal{R}_2)|\mathfrak{g}|$	$O(1)|\mathfrak{g}|$

Table 2 Computational time for ABE schemes

Scheme	Encryption	Decryption
[16]	$O(\mathfrak{R})(T_{Mul} + T_{Pairing})$	$O(\mathfrak{R})(T_{Exp} + T_{Pairing})$
[17]	$O(\mathfrak{R})(T_{Mul} + T_{Pairing})$	$O(\mathfrak{R})(T_{Exp} + T_{Pairing})$
[18]	$O(\mathfrak{R})(T_{Mul} + T_{Pairing})$	$O(\mathfrak{R})(T_{Exp} + T_{Pairing})$
[41]	$O(\mathfrak{R})(T_{Mul} + T_{Pairing})$	$O(1)(T_{Exp} + T_{Pairing})$
[19]	$O(\mathfrak{R})(T_{Mul} + T_{Pairing})$	$O(1)(T_{Exp} + T_{Pairing})$
[42]	$O(\mathfrak{R})(T_{Mul} + T_{Pairing})$	$O(1)(T_{Exp} + T_{Pairing})$
[31]	$O(r_1)T_{Exp} + O(1)T_{Mul}$	$O(r_1)(T_{Exp} + T_{Pairing})$
Our work	$O(1)T_{Exp} + O(r_1)T_{Mul}$	$O(r_1)T_{Mul} + O(1)T_{Pairing}$

Table 3 Feature based on ABE schemes

Scheme	Anonymous key issuing	Constant length ciphertext
[16]	✗	✗
[17]	✗	✗
[18]	✗	✗
[41]	✗	✓
[19]	✗	✓
[31]	✗	✓
[42]	✗	✓
Our scheme	✓	✓

5 Conclusion and Future Work

ABE is the key technology for today's real-time applications like cloud-assisted IoT infrastructure requiring efficient multicast cryptographic need. Researchers have improvised the basic scheme of CP-ABE to add the features of constant-length ciphertext. This research mentions the method to provide the efficient length ciphertext and the anonymous key issuing features. Our scheme works in real-time applications due to communication and computation efficiency as to earlier schemes. At present, the scheme works in a single authority environment. One can work the mentioned scheme to the multi-authority environment with a constant length secret key.

References

1. Shahid N, Aneja S (2017) Internet of things: vision, application areas and research challenges. In: Proceedings of the 2017 international conference on I-SMAC (IoT in Social, Mobile, Analytics and Cloud) (I-SMAC). Palladam, India, pp 583–587

2. Said O, Masud M (2013) Towards internet of things: survey and future vision. Int J Comput Netw 5(1):1–17
3. Mehta R, Sahni J, Khanna K (2018) Internet of things: vision, applications and challenges. Procedia Comput Sci 132:1263–1269
4. Mishra D, Gunasekaran A, Childe SJ, Papadopoulos T, Dubey R, Wamba S (2016) Vision, applications and futurechallenges of internet of things. Ind Manag Data Syst 116(7):1331–1355
5. Miorandi D, Sicari S, De Pellegrini F, Chlamtac I (2012) Internet of things: vision, applications and research challenges. Ad Hoc Netw 10(7):1497–1516
6. Chen C-M, Huang Y, Wang K-H, Kumari S, Wu M-E (2020) A secure authenticated and key exchange schemefor fog computing. Enterp Inf Syst 1–16
7. Xiong H, Kang Z, Chen J, Tao J, Yuan C, Kumari S (2020) Anovel multiserver authentication scheme using proxy resignaturewith scalability and strong user anonymity. IEEE Syst J 1–12
8. Sisinni E, Saifullah A, Han S, Jennehag U, Gidlund M (2018) Industrial internet of things: challenges, opportunities, anddirections. IEEE Trans Industr Inf 14(11):4724–4734
9. Sadeghi AR, Wachsmann C, Waidner M (2015) Securityand privacy challenges in industrial internet of things.In: Proceedings of the 2015 52nd ACM/EDAC/IEEE design automation conference (DAC). San Francisco, CA, USA, pp 1–6
10. Yang F, Wang S, Li J, Liu Z, Sun Q (2014) An overview ofinternet of vehicles. China Commun 11(10):1–15
11. Contreras-Castillo J, Zeadally S, Guerrero-Ibañez JA (2018) Internet of vehicles: architecture, protocols, and security. IEEE Internet Things J 5(5):3701–3709
12. Finkenzeller K (2010) RFID Handbook: fundamentals and applicationsin contactless smart cards, radio frequency identification and near-field communication. Wiley, Hoboken, NJ, USA, p 478
13. Gan G, Lu Z, Jiang J (2011) Internet of things securityanalysis. In: Proceedings of the 2011 international conference on internet technology and applications. Wuhan, China, pp 1–4
14. Hu Z (2011) e research of several key question of internet ofthings. In: Proceedings of the 2011 international conference onintelligence science and information engineering.Wuhan, China, pp 362–365
15. Sahai A, Waters B (2005) Fuzzy identity-based encryption. In: Proceedings of the annual international conference on theory and applications of cryptographic techniques. Aarhus, Denmark, pp 457–473
16. Bethencourt J, Sahai A, Waters B (2007) Ciphertext-policyattribute-based encryption. In: Proceedings of the 2007 IEEE symposium on security and privacy (SP'07). Berkeley, CA, USA, pp 321–334
17. Waters B (2011) Ciphertext-policy attribute-based encryption: anexpressive, efficient, and provably secure realization. In: Proceedings of the international workshop on public key cryptography. Berlin, Germany, pp 53–70
18. Rouselakis Y, Waters B (2012) New constructions and proofmethods for large universe attribute-based encryption. IACRCryptology EPrint Archive 583
19. Guo F, Mu Y, Susilo W, Wong DS, Varadharajan V (2014) CP-ABE with constant-size keys for lightweight devices. IEEE Trans Inf Forensics Secur 9(5):763–771
20. Herranz J, Laguillaumie F, R'afols C (2010) Constant sizeciphertexts in threshold attribute-based encryption. In: Proceedings of the international workshop on public key cryptography. Beijing, China, pp 19–34
21. Susilo W, Yang G, Guo F, Huang Q (2018) Constant-sizeciphertexts in threshold attribute-based encryption withoutdummy attributes. Inf Sci 429:349–360
22. Teng W, Yang G, Xiang Y, Zhang T, Wang D (2015) Attribute-based access control with constant-size ciphertext incloud computing. IEEE Trans Cloud Comput 5(4):617–627
23. Emura K, Miyaji A, Nomura A, Omote K, Soshi M (2009) Aciphertext-policy attribute-based encryption scheme withconstant ciphertext length. In: Bao F, Li H, Wang G (eds) Information security practiceand experience. Springer, Berlin, pp 13–23
24. Fan Y, Liu S, Tan G, Lin X (2019) Cscac: one constant-sizecpabe access control scheme in trusted execution environment. Int J Comput Sci Eng 19(2):162–168

25. Sethia D, Shakya A, Aggarwal R, Bhayana S (2019) Constantsize cp-abe with scalable revocation for resource-constrainediot devices. In: Proceedings of the 2019 IEEE 10th annual ubiquitous computing, electronics and mobile communication conference (UEMCON). New York, NY, USA, pp 0951–0957
26. Zhang Y, Zheng D, Chen X, Li J, Li H (2014) Computationally efficient ciphertext-policy attribute-based encryptionwith constant-size ciphertexts. In: Chow SSM, Liu JK, Hui LCK, Yiu SM (eds) Provable security. Springer International Publishing, Cham, Switzerland, pp 259–273
27. Doshi N, Jinwala DC (2014) Fully secure ciphertext policyattribute-based encryption with constant length ciphertextand faster decryption. Secur Commun Netw 7(11):1988–2002
28. Han J, Yang Y, Liu JK, Li J, Liang K, Shen J (2018) Expressiveattribute-based keyword search with constant-sizeciphertext. Soft Comput 22(15):5163–5177
29. Yang W, Wang R, Guan Z, Wu L, Du X, Guizani M (2020) A lightweight attribute based encryption scheme with constantsize ciphertext for internet of things. In: Proceedings of the ICC 2020–2020 IEEE international conference on communications (ICC). Dublin, Ireland, pp 1–6
30. Green M, Hohenberger S, Waters B (2011) Outsourcing thedecryption of abe ciphertexts. In: Proceedings of the USENIXSecurity symposium. San Francisco, CA, USA
31. Lai J, Deng RH, Guan C, Weng J (2013) Attribute-basedencryption with verifiable outsourced decryption. IEEE Trans Inf Forensics Secur 8(8):1343–1354
32. Lin S, Zhang R, Ma H, Wang M (2015) Revisiting attributebasedencryption with verifiable outsourced decryption. IEEE Trans Inf Forensics Secur 10(10):2119–2130
33. Qin B, Deng RH, Liu S, Ma S (2015) Attribute-based encryptionwith efficient verifiable outsourced decryption. IEEE Trans Inf Forensics Secur 10(7):1384–1393
34. Mao X, Lai J, Mei Q, Chen K, Weng J (2015) Generic andefficient constructions of attribute-based encryption withverifiable outsourced decryption. IEEE Trans Dependable Secure Comput 13(5):533–546
35. Ning J, Cao Z, Dong X, Liang K, Ma H, Wei L (2017) Auditable σ-time outsourced attribute-based encryption foraccess control in cloud computing. IEEE Trans Oninformation Forensics Secur 13(1):94–105
36. Li J, Lin X, Zhang Y, Han J (2017) KSF-OABE: outsourcedattribute-based encryption with keyword search function forcloud storage. IEEE Trans Serv Comput 10(5):715–725
37. Wang H, He D, Han J (2017) VOD-ADAC: anonymousdistributed fine-grained access control protocol with verifiableoutsourced decryption in public cloud. IEEE Trans Onservices Comput 13(3):572–583
38. Xiong H, Zhao Y, Peng L, Zhang H, Yeh K-H (2019) Partially policy-hidden attribute-based broadcast encryptionwith secure delegation in edge computing. Futur GenerationComputer Syst 97:453–461
39. Odelu V, Das AK, Rao YS, Kumari S, Khan MK, Choo K-KR (2017) Pairing-based CP-ABE with constant-sizeciphertexts and secret keys for cloud environment. Comput Stand Interfaces 54:3–9
40. Beimel A (1996) Secure schemes for secret sharing and keydistribution. Technion-Israel Institute of Technology, Haifa, Israel, pp 11–46
41. Lewko A, Waters B (2012) New proof methods for attributebasedencryption: achieving full security through selectivetechniques. In: Proceedings of the annual cryptology conference, pp 180–198. Santa Barbara, CA, USA
42. Zhang Z, Zhang W, Qin Z (2021) Fully constant-size CP-ABE with privacy-preserving outsourced decryption for lightweight devices in cloud-assisted IoT. Secur Commun Netw Article ID 6676862:16

Chapter 10
Parallelizing RP^2-NGSA-II for the Multicriteria Ranking Problem of Sets of Medium-Sized Alternatives Through OpenMP

Gastelum Chavira Diego Alonso

1 Introduction

Today, computers are powerful and allow the execution of algorithms fast. However, there are optimization problems that cannot be addressed reasonably using current computers, which run sequential algorithms that try to find a solution or set of solutions.

According to [4], optimization refers to finding feasible solutions corresponding to extreme values of one or more objectives. Mono-objective optimization problems involve a single objective function; usually, its result is only one solution. It is called multi-objective optimization when an optimization problem is defined with two or more objective functions, and the result includes one or more solutions.

Combinatorial optimization provides a powerful tool for formulating and modeling many optimization problems [25]. Combinatorial optimization is a field of applied mathematics. It combines linear programming, combinatorics techniques, and algorithm theory for solving optimization problems on discrete structures [3]. These problems can be found in most areas of knowledge, and the way to solve them is an open problem, which motivates the creation or adaptation of solution methods for such situations.

Combinatorial optimization is often a realistic and efficient way to deal with many problems. However, multi-objective combinatorial optimization presents specific difficulties. The three main difficulties are (a) the number of efficient solutions, which can be very large; (b) the non-convex characteristic of the feasible set, which requires specific procedures to generate efficient solutions; and (c) the difficulty

G. C. D. Alonso (✉)
Universidad Autonoma de Occidente, 80054 Culiacan SIN, Mexico
e-mail: diego.chavira@uadeo.mx

M. Sahni et al. (eds.), *Novel Developments in Futuristic AI-based Technologies*,
Algorithms for Intelligent Systems, https://doi.org/10.1007/978-981-99-3076-0_10

of developing methods to avoid introducing additional restrictions, which can be retained throughout the procedure [25].

Because combinatorial problems generally have a large solution space, some are intractable, even using current computers. It could happen in multi-objective combinatorial optimization problems. In this type of problem, it is necessary to find the best compromise between two or more objective functions that regularly do not have a single solution but a set of feasible solutions. The complexity of such problems means that the performance of exhaustive search algorithms is often low. The time needed to explore the solution space can be measured in hours, days, and even years; without guaranteeing optimal results. Therefore, using heuristic or metaheuristic algorithms for this problem is advisable.

A heuristic is a procedure based on rules that try to find good solutions with a "low" computational cost. A heuristic is known as an approximate technique and does not ensure optimality. In the same sense, a metaheuristic is an approach that leads a heuristic to solve a problem by conducting and modifying subordinate heuristic operations to obtain solutions [26].

There are various metaheuristic approaches, such as Evolutionary Algorithms, Simulated Annealing, Taboo Search, Ant Colony paradigms, and Bee Swarms with different variations. Multi-objective optimization problems can be addressed through these approaches, which randomly explore the solution space, using techniques that guide these procedures to find optimal solutions.

The Elitist Non-Dominated Genetic Sorting Algorithm (NSGA-II) is one of the most prominent algorithms for solving multi-objective optimization problems. It is a metaheuristic developed by [5] as an extension of the NSGA proposed by [22]. It can be downloaded from https://www.github.com/darnir/nsga2-Windows. NSGA-II has been used or adapted to address different types of problems: scheduling problem with release times for steel plants [13], Software product line development [6], aluminum sheet hot stamping process [27], Interpretability of classification systems (Lajsasna and Seng 2017), geostationary orbit satellite mission planning (Li et al. 2017), location-inventory-routing humanitarian supply chain [24], image encryption using parallelism [9], portfolio optimization [30], multiple traveling salesman problem [29], cancer-associated modules from microRNA [1], trend forecasting and trading in Forex markets [20], Flow Shop Scheduling with Missing Operations [18], worker selection scheme in crowdsourced platforms [28], weight setting problems in traffic engineering [16], and additional problems published in more than 9,000 articles in journals from 2000 to 2022 included in the Scopus database.

In 2016, NSGA-II was adapted to solve the problem of the multicriteria ordering of a set of medium-sized alternatives [11] named RP^2-NSGA-II. For RP^2-NSGA-II and virtually any NSGA-II-based approach, parallel computing elements can be incorporated to reduce the execution time or increase the number of solutions by allowing the search for solutions with more combinations, the latter with runtime like running it without parallelism.

Regardless of the approach used to run a program in parallel, there is a requirement that there is no dependency on data or execution sequence within the program. Therefore, the sections that meet this requirement must be identified in the program's source code. In this sense, this work aims to present the incorporation of parallelism in any algorithm based on NSGA-II, exemplifying it in RP^2-NSGA-II; it is to know the execution time when evaluating in parallel the solutions obtained and the feasibility of implementing parallelism in the said algorithm. The rest of the paper is organized as follows: Sect. 2 introduces parallel computing. Section 3 is intended to present the general features of the original NSGA-II and RP^2-NSGA-II and how parallelism can be added to these implementations. Section 4 is intended to present the performance tests of RP^2-NSGA-II in its parallel version. Section 5 is reserved for results, and Sect. 6 is for conclusions and future work.

2 Parallel Computing

Parallel computing is a computer architecture that allows processes to run simultaneously. Using parallel computing, execution regularly decreases the time required for a specific task; however, it is not always possible. Operations executed on a computer are carried out through processes, which is the instantiation of a running program, and, through them, programs can be executed in parallel. On the other hand, threads allow a program to do more than one thing simultaneously. Like processes, threads can run simultaneously in the kernel of the computer's operating system, scheduling them asynchronously and interrupting each thread so that other threads run [21]. Through the kernel, the operating system has full access to all hardware and can execute any instruction the machine can execute [23].

In parallel computing, there are several ways to classify high-performance architectures, the most common being Multicomputer and Multiprocessor. In the first one, each processor has direct access to its local memory and only to this. On the other hand, a multiprocessor architecture is a computer with more than one processor and where each of them shares the memory with the rest. In this type of architecture, communication between processes is done through shared memory [17]. Multiprocessor architecture is also implemented at the level of the multiple cores of the different types of processors.

Alternatives to implement parallel computing include Application Programming Interface (API) from a specialized hardware vendor, the Portable Operating System Interface (Phreads POSIX) [8], the Open Message Passing Interface (Open MPI) for multicomputer processing [14], and the Open Multiprocessor Processing (OpenMP) for computers with multiple processors or multiple cores. Likewise, a combination of OpenMPI and OpenMP is possible to implement. Regarding the alternatives, OpenMPI and Pthreads POSIX, [2] describe OpenMP.

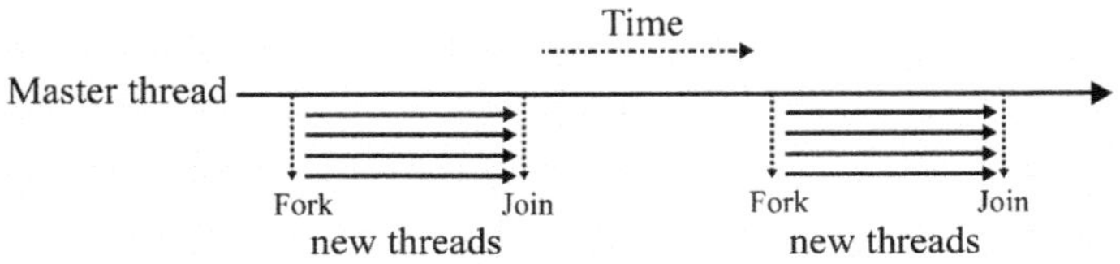

Fig. 1 Shared memory model using the Fork/Join parallelism approach

2.1 Open Parallel Multiprocessing Using OpenMP

When a program is executed, the operating system designates a processor core for that task, even if the computer equipment has a processor with multiple cores. It means that the capacity of the computer is not exploited. A multiprocessor parallel computing approach could solve this situation.

As mentioned, an alternative to implementing multiprocessor parallelism is using OpenMP, which has become a shared memory standard. OpenMP is a set of compiler directives and a library of supporting functions that work in conjunction with the standard Fortran and C/C++ languages [17].

A shared memory program that implements parallelism is based on the Fork/Join approach. When you run a program, a main thread is created. This main thread successively executes the instructions of the program. If a parallel region is defined in a program, the main thread creates or forks other threads to execute the operations of that region in parallel.

The created threads are paused or killed when the parallel region is finished. After that, the main thread takes control of the program's sequence (Join). A sequential program is a special case of a parallel shared memory program when it does not perform Fork/Join. Figure 1 shows the Fork/Join execution model in parallel computing using shared memory.

2.2 Parallelizing for Loops with OpenMP

The directives that include OpenMP allow you to run cyclic structures in parallel; specifically, it has one exclusively for the For loop, the #pragma parallel for directive. A compiler directive in C or C++ is called a pragma. The word pragma is derived from pragmatic information. A pragma is a way of communicating information to the compiler, which can help the compiler optimize the program [17]. The way to parallelize a For loop is by placing #pragma omp parallel for before defining the For loop with its instruction body. The following program shows the parallelization of a For loop in C using OpenMP.

```
Input: None
Output: Values of variable i followed by thread number
File name: parallelProgram.c
1.  #include <stdio.h>
2.  #include <omp.h>
3.  int main (){
4.  int i;
5.  #pragma omp parallel for
6.  for (i= 0; i<10; i++)
7.       printf("value of i: %d, number of thread: %d\n",
         i, omp_get_thread_num());
8.  return 0;
9.  }
```

With the #pragma omp parallel for directive, lines 6 and 7 of the previous program will be executed in parallel. To be parallelized, a For loop should not contain instructions in the body that cause the cycle to end early, such as break, return, exit, and goto. Likewise, there should be no dependencies between the data. All programs that use OpenMP must include the omp.h header at the beginning of the file, as in line 2. The omp_get_thread_num() function on line 7 allows getting the thread number who performed the task.

The parallelProgram.c file must be compiled by adding the-fopenmp option on the command line to create the executable or binary file. Thus, to compile to parallelProgram.c using the gcc compiler (GNU C Compiler), the following instruction is executed from the command line: gcc-fopenmp parallelProgram.c -o parallelProgram.

Once a program containing OpenMP has been compiled, it is necessary to indicate the number of threads desired during its execution. It can be done by the OMP_NUM_THREADS environment variable, which is assigned the number of threads using the export command on Unix/GNU-Linux and set on Windows. For instance, in GNU/Linux, to assign three threads, export OMP_NUM_THREADS = 3 is executed without adding space between the = sign and the desired number of threads. Subsequently, parallelProgram is run like any other executable in GNU/Linux from the command line:./parallelProgram. A result of the execution of the parallelProgram is shown in Fig. 2. There, the For loop was executed in parallel and asynchronously in the three threads defined in the variable OMP_NUM_THREADS; the printf function found on line 7 in the previous program was not executed sequentially.

3 General Characteristics of NSGA-II and RP^2-NSGA-II

At the beginning of its execution, NSGA-II creates a set of solutions to the problem in question at random, called the initial population. Subsequently, it generates new solutions (descendant population) through selection techniques and crossover and

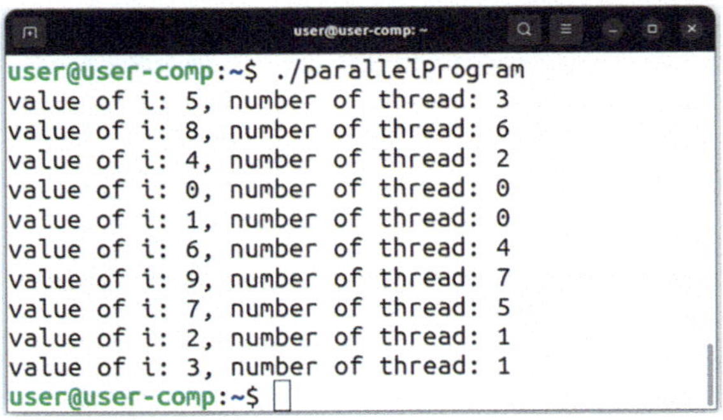

Fig. 2 Result of running the programParallel binary

mutation genetic operators. A mechanism evaluates these solutions called crowding distance. After that, an ordered set of non-dominated solutions is created. Two solutions, sol_1 and sol_2, are non-dominated if sol_1 is better by one or more values than sol_2 in the objective functions and vice versa. The solution sol_2 is dominated if solution sol_1 is better than it in all the objective functions of the problem under consideration.

NSGA-II uses elitism as a method of population replacement and an explicit strategy to preserve the diversity of that population. It implies assigning a diversity level or rank to each individual that belongs to the identical non-domination front. Therefore, each individual in the population has a non-domination position. The individuals with the best rank form a non-domination front called the Pareto front. The individuals within the Pareto front are in the least crowded region; thus, they are ranked higher. Then, NSGA-II computes the density of solutions surrounding a particular solution in the population. It is done by using values of each objective function in the crowding distance. NSGA-II executes the operations of selection, crossover, mutation, evaluation, and others in several generations (iterations) defined at the beginning of its execution. In the end, NSGA-II presents solutions that are not necessarily all non-dominated among themselves.

3.1 NSGA-II Input Parameters and File Structure

NSGA-II requires a set of input parameters for its execution. These parameters are, among others, population size, generations, objectives, restrictions, variables, mutation, and crossover operators. The parameters can be entered step by step when executing NSGA-II or through a parameter file from the command line as follows: $ > nsga2r < parameters_file seed_for_random. Only two levels of headings should be numbered. Lower-level headings remain unnumbered; they are formatted as run-in headings.

The implementation of NSGA-II is in C language in a series of files (twenty C files and two headers files). The variables, structures, and functions used in practically all

NSGA-II are declared in global.h. This work used the nsga2r.c and eval.c files. The first is the principal file of NSGA-II, and the second is used to evaluate the solutions (individuals) generated by NSGA-II.

3.2 *The General Operation of the Main Function of NSGA-II*

The general operation of the NSGA-II algorithm is presented below. It should be noted that the part of the NSGA-II where the procedure that evaluates the population is invoked is of interest. The next program shows the structure of the NSGA-II main file.

```
Input: Parameter set
Output: Set of solutions found
File name: parallelProgram.c
1. randomize();
2. initialize_pop(parent_pop);
3. decode_pop(parent_pop);
4. evaluate_pop (parent_pop);
5. assign_rank_and_crowding_distance(parent_pop);
6. for (i=2; i<=ngen; i++){
7.     selection (parent_pop, child_pop);
8.     mutation_pop (child_pop);
9.     decode_pop(child_pop);
10.    evaluate_pop(child_pop);
11.    merge(parent_pop, child_pop, mixed_pop);
12.    fill_nondominated_sort(mixed_pop, parent_pop);
13.    report_pop(parent_pop,fpt4);
14.}
```

In the previous program, parent_pop, child_pop, and mixed_pop are structures that respectively represent the parent and descendant populations and the union of both. The ngen is a variable that indicates the number of generations or iterations that NSGA-II will have. The argument fpt4 is a file pointer to save the results in a file.

When lines 4 and 10 of the previous program are executed, the function evaluate_pop() located in the file eval.c is called. This evaluate_pop() function receives as input an array that represents the population to be evaluated. Inside, this function has a For loop that loops through the array row by row, i.e., individual by individual. On each For loop iteration, the function evaluate_ind() is called, located in eval.c, which receives one individual at a time to be evaluated. The next program shows the structure of the evaluate_pop() function.

```
Input: Population
Output: Set of individuals evaluated
File name: eval.c
void evaluate_pop (population *pop){
int i;
for (i=0; i<popsize; i++)
        evaluate_ind (&(pop->ind[i]));
return;
}
```

: Code of evaluate_ind() function

In this program, *pop is the population to evaluate, which is defined for global access in the global.h file. popsize is the population size entered at the start of the program. pop- > ind[i] is the ith individual of the population.

3.3 RP²-NSGA-II for the Multicriteria Ordering Problem

As mentioned in the introduction, NSGA-II has been used and adapted to address different types of problems, among them, for the treatment of the multicriteria ordering problem of a set of medium-sized alternatives such as the one presented in [11]. This adaptation of the NSGA-II algorithm consists of partitioning a set of medium-sized decision alternatives into classes and seeks to derive a ranking of classes with a partial or full order structure, where the decision alternatives within a class are indifferent to each other. This implementation is called RP^2-NSGA-II and receives a fuzzy outranking relation as input.

In [11], the multicriteria ordering problem was formulated as a multi-objective combinatorial optimization problem, where the final result consists in obtaining an ordered partition of a set of alternatives $A = \{a_1, a_2, .., a_m\}$, evaluated by a set of criteria $G = \{g_1, g_2, .., g_n\}$, taking into account the preferences of a decision maker, represented in a fuzzy outranking relation. Such a partition comprises classes or groups of alternatives indifferent to each other. Defining the characteristics of fuzzy outranking relations and how to model decision-maker preferences is beyond this work's scope. Still, readers can see [7, 19] for an explanation of them.

RP^2-NSGA-II tries to optimize three objective functions simultaneously, one to be maximized and two to be minimized. RP^2-NSGA-II considers the association of a fuzzy outranking relation F_{OR} with a cutting level to define a crisp outranking relation C_{OR}. It allows for deriving a preference structure that models the decision-maker preferences. RP^2-NSGA-II attempts to obtain cutting level values near one. It is called the maximum cutting-level objective.

The approach attempts to assign only indifferent alternatives inside each class. It is known as the min-cut objective. Finally, RP^2-NSGA-II seeks to minimize the number of discrepancies between the C_{OR} and the obtained ranking of alternatives. It

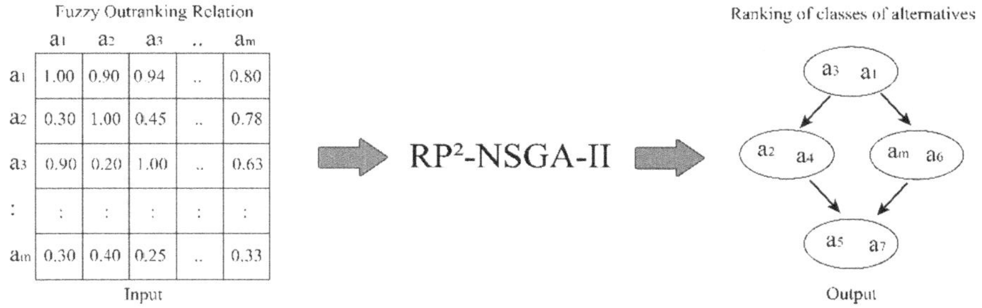

Fig. 3 Example of a multicriteria ranking of alternatives obtained by RP^2-NSGA-II

is called the minimum pairwise preference disagreement objective. The description of the model, the implementation, and the operation of RP^2-NSGA-II are presented in [11]. Figure 3 shows an output generated by that algorithm.

In a multicriteria ordering problem of medium-sized sets of alternatives, it is recognized that when the number of alternatives and classes increases, the space of possible solutions grows exponentially. Thus, the required time to find non-dominated solutions grows too. Therefore, it is a complex problem to be solved with exhaustive search techniques and even metaheuristics.

3.4 *Input Parameters for RP^2-NSGA-II*

For its execution, RP^2-NSGA-II requires practically the same input data as NSGA-II, except for a file containing a fuzzy outranking relation. The way to introduce these parameters is kept the same as NSGA-II; therefore, they can be supplied step by step when executing the algorithm or through the command line, as follows: $ > nsga2r < parameters_file seed_for_random input_data_file.

3.5 *Problem Definition to Parallelize RP^2-NSGA-II*

A parallel version of RP^2-NSGA-II can be created using OpenMP to evaluate the populations. Thus, the problem of evaluating the populations in parallel was established as follows.

Let $A = \{a_1, a_2, .., a_m\}$ be a finite set of individuals from a population of the RP^2-NSGA-II multi-objective evolutionary algorithm. Let $|A|$ be the number of individuals in the population of RP^2-NSGA-II. Let $F = \{f_1, f_2, f_3\}$ be the set of the three objective functions defined by [11] to evaluate each individual in A. Let n be the generations that RP^2-NSGA-II will use to search the solutions. The problem is how to evaluate A using F on the n generations of RP^2-NSGA-II in a parallel form through

OpenMP to use the cores of a computer's central processing unit, thereby reducing the evaluation time of individuals of A in the n generations.

4 Methodology

This contribution focuses on incorporating Open MP into the RP^2-NSGA-II algorithm to evaluate individuals in a population in a parallel form. In the same way, it is focused on knowing the performance of this using different execution threads in an Intel i7 processor with eight cores. Likewise, the work centers on knowing the number of threads with which the algorithm performed better. This RP^2-NSGA-II algorithm with OpenMP is called here as RP^2-NSGA-II_Parallel. Likewise, it included the tests of this new parallel version of RP^2-NSGA-II.

For the incorporation of OpenMP in RP^2-NSGA-II, an identification process was carried out where it is possible to include the parallel section(s). It was identified that the eval.c file has the evaluate_pop() function implemented, which evaluates the individuals of a population. This function is called from the RP^2-NSGA-II main file.

The evaluation of individuals in a population in parallel needs some OpenMP directives in the eval.c file. These directives are shown in the next program in lines 1 and 4 of the eval.c file.

```
Input: Population
Output: Set of individuals evaluated in parallel
File name: eval.c
1. #include <omp.h>
2. void evaluate_pop (population *pop) {
3. int i;
4. #pragma omp parallel for
5. for (i=0; i<popsize; i++)
6.     evaluate_ind (&(pop->ind[i]));
7. return; }
:
```

4.1 *Empirical Test Design*

Four empirical tests were designed to evaluate the performance of RP^2-NSGA-II_Parallel. The first was following the experimental design shown in Table 1. This test aimed to know the performance time of RP^2-NSGA-II_Parallel with different sizes of fuzzy outranking relations that structurally contain different number of classes using different number of threads. For this first test, the population included 100

Table 1 Experimental design for the first empirical test

Variable	Values
V_1: Quantity of classes	3, 5, 7, 9, 11
V_2: Quantity of threads	1, 2, 3, 4, 5, 6, 7, 8
V_3: Quantity of alternatives	200, 400

Table 2 Experimental design for the second empirical test

Variable	Values
V_1: Quantity of classes	3, 5, 7, 9, 11
V_2: Quantity of threads	1, 2, 3, 4, 5, 6, 7, 8

individuals and 5,000 generations. Crossover and mutation probabilities were set at 90 and 10%, respectively.

The second empirical test aimed to know the performance time of the algorithm with fuzzy outranking relations of 200 × 200 that structurally contains different number of classes, also using different number of threads. The population included 100 individuals and 10,000 generations. Crossover and mutation probabilities were defined as 90 and 10%, respectively. Table 2 shows the experimental design of this second test.

In the same sense, the third test aimed to find out if the proportion of reduction in execution times was maintained with what was obtained in the two previous empirical tests. This test used a fuzzy outranking relation of 400 × 400 with three classes, 200 individuals, 10,000 generations, and crossover and mutation probabilities of 90 and 10%, respectively, but only for 1 and 8 threads.

Finally, the population initialization process of RP^2-NSGA-II_Parallel was parallelized. This process uses the k-means algorithm to generate the first half of the population and the Minimum Spanning Tree (MST) for the rest. This test aimed to know the time reduction for this initialization process, which is different from the random initialization process of NSGA-II.

In all tests, RP^2-NSGA-II_Parallel was provided with fuzzy outranking relations via plain text files. Each relation was a 200 × 200 or 400 × 400 matrix with fuzzy values [0, 1] representing a pairwise comparison of 200 or 400 alternatives. These relations were generated using the instance generator presented in the next algorithm.

```
Input: num_alternatives NA, num_classes NC, cutting-level
Output: FuzzyOutrankingRelation, CrispOutrankingRelation
1. Begin.
2. Randomly create an Nc×Nc matrix to represent a
   partially ordered set Adj_matrix.
3. Randomly fill an array Alt_array of size NA so that
   each alternative belongs to a class.
4. Randomly define a cutting_level → [0.60,0.75].
5. Create two NA×NA matrices to represent a Fuzzy
   Outranking Relation and a Crisp Outranking Relation
   using the Alt_array, Adj_matrix, and
   cutting_level values.
6. End.
```

As mentioned, for the empirical tests, the number of threads used varied from 1 to 8. Running RP^2-NSGA-II_Parallel with OMP_NUM_THREADS = 1 is equivalent to using a non-parallel algorithm because it runs with a single thread on the computer's CPU. The computer on which the tests were carried out corresponded to an Alienware R2 15 (2.40 GHz, 8-cores), 24 GB of base memory (DDR3-1066 GHz) and M.2 1 TB 3.5 MB/s storage device, and GNU Ubuntu operating system 22.04. This computer was used exclusively to carry out the evaluation. The number of threads was made in a GNU-Linux terminal through the export command and the environment variable OMP_NUM_THREAD, as presented in Sect. 2.2.

The Ubuntu time command was used to evaluate the performance of RP^2-NSGA-II_Parallel. This command returns the execution time used by a program. The invocation of this command and RP^2-NSGA-II_Parallel was carried out in a terminal of the form: time./nsga2-rankingfinal < file_of_parameters seed_for_random file_of_input_data.

When the time command ends, a message with program time statistics is obtained. Those statistics are displayed as real, user, and sys. The first one corresponds to the real-time elapsed between the invocation and completion of the command. The second is related to the processor's user time. The third refers to the system time of the processor [15]. Figure 4 shows the output of running nsga2-rankingfinal_parallel for a fuzzy outranking relation of 200 × 200 with three classes stored in the file sigma3_200_0. The parameters of that execution were configured in the file settings.in with 100 individuals, 200 chromosomes per individual, 5000 generations, 90% crossover probability, and 10% probability of mutation. For the execution of the algorithm, a value of 0.2 was defined as a randomized seed, and OMP_NUM_THREADS = 1 to use only one thread.

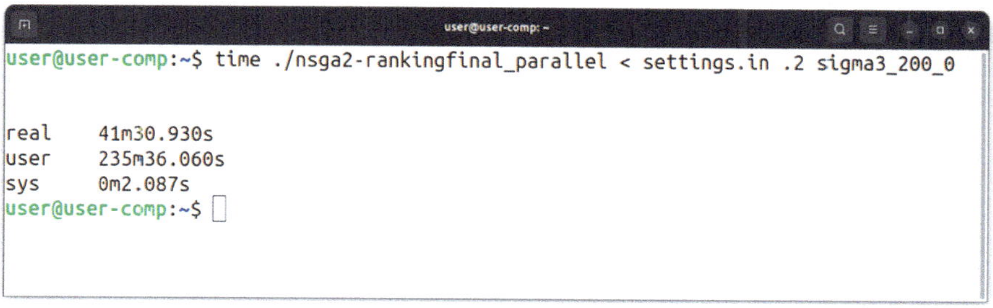

Fig. 4 RP^2-NSGA-II_Parallel execution times

5 Results

The results of the first empirical test, where the population consisted of 100 individuals, 200 and 400 chromosomes per individual, 5,000 generations, 90% crossover probability, and 10% probability of mutation, are shown in Table 3. For the fuzzy outranking relations of 200 alternatives, RP^2-NSGA-II_Parallel had the best performance in real time with seven classes and the worst with 9 and 11 classes, regardless of the number of threads. Likewise, it is observed that when the total of threads was increased, the real execution time of RP^2-NSGA-II_Parallel decreased considerably. On the other hand, for fuzzy outranking relations of 400 alternatives, RP^2-NSGA-II_Parallel had the best real-time performance for 11 classes and the worst performance with five classes. Like the previous results, when the total of threads was increased, the real-time execution of RP^2-NSGA-II_Parallel decreased considerably.

On the other hand, Table 4 shows the percentages of time reduction between using only one thread or more threads for 200 and 400 alternatives. Also, it shows the average percentage of such reduction times. There, it is possible to observe the benefits of using a parallel approach, which takes advantage of the capabilities of the computer's processor for RP^2-NSGA-II_Parallel or any other that can be parallelized.

Likewise, the results in Table 5 correspond to the second empirical test of RP^2-NSGA-II_Parallel. There, different fuzzy outranking relations of 200×200 were supplied, with a configuration fixed in population size of 100 individuals, 200 chromosomes, 10,000 generations, crossover probability 90%, mutation probability 10%, for 1, 2, 3, 4, 5, 6, 7, and 8 threads.

Based on Table 5, the real-time differences of each execution of RP^2-NSGA-II_Parallel can be obtained. The percentages correspond to the real-time difference between using one thread and 2, 3, 4, 5, 6, 7, and 8 threads. The real-time differences in using from 1 to 8 threads can be observed in Table 6. This Table indicates, for instance, that if eight threads are used instead of a single thread, real-time reductions of −82, −73, −71, −72, −74, and −74% for 11, 9, 7, 5, and 3 classes, respectively. On average, 74% was the real-time reduction between using a single thread and eight threads.

Table 3 Real time used by RP^2-NSGA-II_Parallel in the first test

Alternatives	Threads	3 classes	5 classes	7 classes	9 classes	11 classes
200	1	5m18.73 s	4m35.96 s	4m50.06 s	5m30.96 s	6m55.05 s
	2	3m20.29 s	2m52.96 s	2m53.83 s	3m34.38 s	4m3.14 s
	3	2m40.09 s	2m16.32 s	2m15.20 s	2m53.41 s	3m2.77 s
	4	2m18.94 s	1m57.34 s	1m53.90 s	2m33.09 s	2m34.17 s
	5	2m12.84 s	1m52.17 s	1m49.69 s	2m25.10 s	2m23.46 s
	6	2m5.89 s	1m45.96 s	1m42.12 s	2m19.14 s	2m14.23 s
	7	2m0.24 s	1m40.54 s	1m36.89 s	2m11.33 s	2m6.63 s
	8	1m54.47 s	1m36.79 s	1m31.38 s	2m7.36 s	1m58.14 s
400	1	82m42.9 s	89m38.4 s	61m53.6 s	33m12.7 s	27m26.6 s
	2	46m9.6 s	50m20.9 s	34m31.9 s	21m14.1 s	17m56.2 s
	3	33m45.9 s	36m27.2 s	25m13.1 s	16m57.4 s	14m48.7 s
	4	27m16.9 s	29m32.2 s	20m25.8 s	15m7.2 s	13m0.4 s
	5	24m43.9 s	26m42.9 s	18m37.3 s	14m20.1 s	12m10.4 s
	6	22m46.5 s	24m22.3 s	17m2.7 s	13m11.7 s	11m33.7 s
	7	21m6.7 s	22m41.1 s	15m49.6 s	12m52.5 s	11m3.9 s
	8	19m14.0 s	20m42.6 s	14m36.4 s	12m8.7 s	10m37.1 s

Table 4 Time differences between using one thread and more threads with RP^2-NSGA-II_Parallel for different number of alternatives and classes

Alternatives	Threads	3 classes (%)	5 classes (%)	7 classes (%)	9 classes (%)	11 classes (%)	Average (%)
200	2	−38	−42	−44	−37	−34	−39
	3	−54	−50	−52	−52	−51	−52
	4	−58	−64	−66	−56	−64	−62
	5	−59	−65	−67	−58	−66	−63
	6	−52	−67	−68	−59	−67	−63
	7	−61	−68	−70	−60	−60	−64
	8	−70	−69	−71	−49	−76	−67
400	2	−43	−44	−44	−36	−36	−41
	3	−59	−59	−59	−50	−47	−55
	4	−67	−67	−67	−53	−52	−61
	5	−70	−70	−70	−57	−56	−65
	6	−73	−73	−72	−60	−58	−67
	7	−74	−75	−75	−62	−59	−69
	8	−77	−77	−77	−61	−62	−71

Table 5 Real time used by RP^2-NSGA-II_Parallel in the second empirical test

Threads	3 classes	5 classes	7 classes	9 classes	11 classes
1	169.34 m	128.49 m	121.59 m	145.36 m	216.26 m
2	96.18 m	74.47 m	72.25 m	83.51 m	124.70 m
3	71.11 m	56.38 m	55.14 m	63.20 m	92.10 m
4	58.24 m	47.60 m	46.15 m	52.20 m	75.37 m
5	54.00 m	44.11 m	43.14 m	48.36 m	69.40 m
6	50.24 m	40.43 m	40.11 m	44.57 m	64.30 m
7	47.00 m	38.15 m	37.52 m	41.55 m	41.59 m
8	44.48 m	36.60 m	35.59 m	39.30 m	39.25 m

Table 6 Differences in time using one thread versus several threads in RP^2-NSGA-II_Parallel

Threads	3 classes	5 classes	7 classes	9 classes	11 classes	Average
1	–	–	–	–	–	–
2	−43%	−42%	−41%	−43%	−42%	−42%
3	−58%	−56%	−55%	−57%	−57%	−57%
4	−66%	−63%	−62%	−64%	−65%	−64%
5	−68%	−66%	−65%	−67%	−68%	−67%
6	−70%	−69%	−67%	−69%	−70%	−69%
7	−72%	−70%	−69%	−71%	−81%	−73%
8	−74%	−72%	−71%	−73%	−82%	−74%

The results of the third test, where a fuzzy outranking relationship of 400×400 was provided with three classes, 200 individuals, 10,000 generations, 90% crossover probability, 10% mutation probability, and eight threads, indicate that for the execution of RP^2-NSGA-II_Parallel with only thread, 7523 min (5.22 days) were required. From this time, the population initialization process, which was not parallelized, required 21.5 min, and each generation, on average, required 45 s. On the other hand, 1521 min (1.06 days) were required for eight threads, where each generation, on average, required 9 s. Using eight threads instead of one only reduced 80% of the required real time.

Regarding the fourth test with the population initialization process, which before being parallelized lasted approximately 21.5 min, it was found this process required 5.5 min for execution with eight threads, a fuzzy outranking relation of 400×400 with three classes, 200 individuals, 10,000 generations, 90% crossover probability, and 10% mutation probability. This reduction represented a decreased time of 75%, approximately.

6 Concluding Remarks and Future Perspectives

Many problems in the different knowledge areas require high computational resources and execution time. It is presented in various combinatorial optimization problems. Thus, a parallel technique has been used to defeat this in the multicriteria ranking problem of sets of medium-sized alternatives. The implementation of parallelism to the RP^2-NSGA-II multicriteria ordering algorithm was a multicore approach through OpenMP. Using OpenMP only requires the definition of the section to be parallelized, avoiding explicitly creating the threads, making them run, synchronizing, and killing them, as is required with POSIX.

The incorporation of OpenMP to RP^2-NSGA-II to evaluate the individuals of the population in a parallel form reduces the execution times significantly for eight threads concerning only one. The difference of using two execution threads instead of one has a time reduction of at least 41%.

Using OpenMP for RP^2-NSGA-II population initialization and evaluation in parallel is practical and convenient. It is because it is only necessary to evaluate the feasibility of including the OpenMP directives in the RP^2-NSGA-II code and identify where it is most convenient to place them. However, increasing the number of threads does not guarantee that RP^2-NSGA-II_Parallel or another program will perform better since this also increases the synchronization time between the threads, which could cause more CPU time to be spent on that task and may degrade the time performance of RP^2-NSGA-II_Parallel.

In evolutionary algorithms, population sizes of between 40 and 80 individuals are regularly defined for reasons of evaluation time. It implies reducing the number of solutions to be evaluated. With the addition of a parallel approach such as OpenMP, the population size could be increased using the same required execution time without its use. The implementation of OpenMP in RP^2-NSGA-II can be done the same way in all the methods where the original NSGA-II is the base.

Experimental results have shown that the approach reduces the computational execution time compared to using just one core of the CPU. Despite the experimental results, it is necessary to analyze whether it is possible to include new parallel sections in crossover and mutation operators of RP_2-NSGA-II_Parallel. It, in theory, would reduce its total execution time. Optionally, a parallel version of RP^2-NSGA-II can be made using the message-passing interface OpenMPI, which allows parallel execution using multiple computers. However, this version requires a computer cluster infrastructure for its execution.

References

1. Biswas P, Mukhopadhyay A (2020) Identifying cancer-associated modules from microRNA co-expression networks: a multi-objective evolutionary approach. Soft Comput 24:17365–17376. https://doi.org/10.1007/s00500-020-05025-0

2. Chapman B, Jost G, Van Der Pas R (2008) Using OpenMP: portable shared memory parallel programming (Scientific and Engineering Computation). The MIT Press, USA
3. Cook W, Cunningham W, Pulleyblank W, Schrijver A (1998) Combinatorial optimization. Wiley, USA. ISBN: 978-0471558941
4. Deb K (2001) Multi-objective optimization using evolutionary algorithms. Wiley, England
5. Deb K, Pratap A, Agarwal S, Meyarivan T (2002) A fast and elitist multi-objective genetic algorithm: NSGA-II. IEEE Trans Evol Comput 6(2):182–197. https://doi.org/10.1109/4235.996017
6. dos Santos Neto PDA, Britto R, Rabêlo RDAL, Cruz JJDA, Lira WAL (2016) A hybrid approach to suggest software product line portfolios. Appl Soft Comput J 49:1243–1255. https://doi.org/10.1016/j.asoc.2016.08.024
7. Fodor J, Roubens M (1994) Fuzzy preference modeling and multicriteria decision support. Kluwer Academic Publishers, Dordrecht
8. IEEE (2022) IEEE 9945–1–1996 portable operating system interface (POSIX(TM))-Part 1: System Application Program Interface (API) [C Language] ISO/IEC 9945–1: 1996. https://standards.ieee.org/ieee/9945-1/2366/.
9. Kaur M, Kumar V (2018) Parallel non-dominated sorting genetic algorithm-II-based image encryption technique. Imaging Sci J 66(8):453–462. https://doi.org/10.1080/13682199.2018.1505327
10. Lahsasna A, Seng WC (2017) An improved genetic-fuzzy system for classification and data analysis. Expert Syst Appl 83:49–62. https://doi.org/10.1016/j.eswa.2017.04.022
11. Leyva J, Solano J, Garcia J, Gastelum D (2016) Exploitation of a medium-sized fuzzy outranking relation based on multi-objective evolutionary algorithms to derive a ranking. Int J Comput Intell Syst 9(4):745–764. https://doi.org/10.1080/18756891.2016.1204122
12. Li J, Zhang S, Liu X, He R (2017) Multi-objective evolutionary optimization for geostationary orbit satellite mission planning. J Syst Eng Electron 28(5):934–945. https://doi.org/10.21629/JSEE.2017.05.11
13. Long J, Zheng Z, Gao X, Pardalos PM (2016) A hybrid multi-objective evolutionary algorithm based on NSGA-II for practical scheduling with release times in steel plants. J Oper Res Soc 67(9):1184–1199. https://doi.org/10.1057/jors.2016.17
14. MPI (1994) MPI: a message passing interface. Int J Supercomput Appl 8(3/4). https://doi.org/10.1145/169627.169855
15. OpenGroup (2018) Time–time simple command. The Open Group Base Specifications Issue 7, 2018 edition. IEEE Std 1003.1–2017 (Revision of IEEE Std 1003.1–2008). https://pubs.opengroup.org/onlinepubs/9699919799/utilities/time.html
16. Pereira V, Sousa P, Rocha M (2022) A comparison of multi-objective optimization algorithms for weight setting problems in traffic engineering. Nat Comput 21(3):507–522. https://doi.org/10.1007/s11047-020-09807-1
17. Quinn MJ (2004) Parallel programming in C with MPI and OpenMP. McGraw Hill, New York, USA
18. Rossit DG, Rossit DA, Nesmachnow S (2021) Explicit multi-objective evolutionary algorithms for flow shop scheduling with missing operations. Program Comput Soft 47:615–630. https://doi.org/10.1134/S0361768821080223
19. Roy B (1991) The outranking approach and the foundations of electre methods. Theor Decis 31:49–73. https://doi.org/10.1007/BF00134132
20. Sadeghi A, Daneshvar A, Zaj M (2021) Combined ensemble multi-class SVM and fuzzy NSGA-II for trend forecasting and trading in Forex markets. Expert Syst Appl (185). https://doi.org/10.1016/j.eswa.2021.115566
21. Samuel A, Oldham J, Mitchell M (2001) Advanced linux programming. Sams, USA
22. Srinivas N, Deb K (1994) Multi-objective function optimization using nondominated sorting genetic algorithms. Evol Comput 2(3):221–248
23. Tanenbaum AS (2009) Sistemas operativos modernos. 3ra. edición. Pearson Educación, México
24. Tavana M, Abtahi A, Di Caprio D, Hashemi R, Yousefi-Zenouz R (2018) An integrated location-inventory-routing humanitarian supply chain network with pre- and post-disaster management considerations. Socioecon Plann Sci 64:21–37. https://doi.org/10.1016/j.seps.2017.12.004

25. Teghem J (2009) Multi-objective Combinatorial Optimization –MOCO. In: Floudas Ch, Pardalos P (eds) Encyclopedia of optimization, 2nd ed. Springer Science + Business Media, LLC, New York, USA, pp 2434–2442
26. Voss S (2009) Metaheuristics. In: Floudas Ch, Pardalos, P (eds) Encyclopedia of optimization, 2nd ed. Springer Science + Business Media, LLC, New York, USA, pp. 2434–2442
27. Xiao W, Wang B, Zhou J, Ma W, Yang L (2016) Optimization of aluminum sheet hot stamping process using a multi-objective stochastic approach. Eng Optim 48(12):2173–2189. https://doi.org/10.1080/0305215X.2016.1163483
28. Yadav A, Mishra S, Sairam AS (2022) A multi-objective worker selection scheme in crowd-sourced platforms using NSGA-II. Expert Syst Appl 201. https://doi.org/10.1016/j.eswa.2022.116991
29. Yang S, Shao Y, Zhang K (2019) An effective method for solving multiple traveling salesman problem based on NSGA-II. Syst Sci Control Eng 7(2):108–116. https://doi.org/10.1080/21642583.2019.1674220
30. Zhou Z, Liu X, Xiao H, Wu S, Liu Y (2019) A DEA-based MOEA/D algorithm for portfolio optimization. Cluster Comput 22(Suppl 6):14477–14486. https://doi.org/10.1007/s10586-018-2316-7

Chapter 11
The Financial Portfolio Selection Using the Multiple Criteria Hierarchical Process and the Markowitz Model

Maria Bernal, Deyanira Velázquez, Pavel Anselmo Alvarez, Manuel Muñoz-Palma, and Ernesto León-Castro

1 Introduction

The portfolio selection problem consists of determining an investment portfolio that maximises returns given a certain risk level or minimises the risk given a specific return level [10]. It is related to selecting stocks by their characteristics and portfolios with the stocks chosen to obtain a return, seeking to make the maximum return. The portfolio selection problem arises because the performance depends on many factors beyond the investor's control, which is why a low-risk decision problem is generated.

It has been pointed out that each investor has their profile, and their preferences regarding the relative importance of the decision criteria for choosing the assets that make up a portfolio [14]. It is a determining aspect to achieve efficient portfolios

M. Bernal (✉)
Universidad Tecnologica de Culiacan, Carr. Culiacan-Imala, Km 2, 80014 Los Angeles, Culiacan Rosales, Mexico
e-mail: maria.bernal.maba@gmail.com

D. Velázquez · P. A. Alvarez
Department of Management and Economic Sciences, Universidad Autónoma de Occidente, 80020 Culiacan, Mexico
e-mail: laura.velazquez@uadeo.mx

P. A. Alvarez
e-mail: pavel.alvarez@uadeo.mx

M. Muñoz-Palma
Management Department, Universidad de Sonora, Hermosillo, Sonora, Mexico
e-mail: manuel.munoz@unison.mx

E. León-Castro
Business School, Tecnológico de Monterrey, Campus Sinaloa, Blvd. Pedro Infante 3773 Pte, Recursos Hidráulicos, Culiacán Rosales, Sin 80100, México
e-mail: ernesto.leon@tec.mx

M. Sahni et al. (eds.), *Novel Developments in Futuristic AI-based Technologies*, Algorithms for Intelligent Systems, https://doi.org/10.1007/978-981-99-3076-0_11

for an investor in question. The efficient portfolio is the highest return, given a level of risk, or the lowest risk given an expected return. Therefore, the diversification strategy should be linked with investor risk behaviour [26].

The portfolio selection problem can be seen as a process divided into two stages. In the first stage, attractive shares for the investor are selected from the existing universe of shares by evaluating the stock performance based on various criteria representing the investor's preferences. In the second stage, the portfolio is built to distribute the available capital budget among the previously selected actions. In such a way, the budget proportion for each action meets its desired risk and return conditions [11, 19, 26, 32, 33]. Therefore, it is considered that both stages are essential to define a portfolio. The first stage helps evaluate the actions and selects the best actions, thus reducing the size of actions to make up the portfolio. The second stage allows the investor to specify the amount of capital budget to invest in each stock selected in the former stage.

The evaluation of shares in the Stock Exchange is through many indicators. For this kind of problem, a multidimensional approach as multicriteria decision-making (MCDM) is necessary to consider many criteria at the same time to compare the shares. MCDM provides various methods that support the indicators and evaluate the shares. Due to the limitation of the human capacity to handle that much information, some methods considering various criteria simultaneously are needed. In the MCDM methods, some approaches based on multiattribute utility, reference level, or outranking are presented. The present work's applied method is the ELimination Et Choix Traduisant la REalité (ELECTRE). The ELECTRE-III is a ranking method that supports the decision-maker in the evaluation and analysis process. It is adequate for the stock selection problem, because of its useful adaptation for different attribute values and the flexibility to adapt the decision-maker preference.

In this research, the portfolio selection process is carried out in the two stages of the Mexican Stock Exchange. In the first stage, the Multiple Criteria Hierarchical Process (MCHP) is applied to evaluate stocks with various financial performance ratios. They are book values, and market values, used as decision criteria in the aggregation model, are analysed through a hierarchical structure. This method allows the generation of a ranking of companies by subgroups of criteria and a global ranking. MCHP allows the identification of the stocks with the best performance.

In the second stage, the Markowitz model is used to build portfolios with the stocks located in the first positions of the global ranking generated by MCHP. The evaluation on the first stage with MCHP allows the minimisation of risk (represented by the variance), given a specific return level (represented by the average). An optimisation process is used to obtain various portfolios with different returns that minimise the risk, forming the so-called efficient frontier.

The structure of the paper is the following. In Sect. 2, the literature review of similar works that use methodologies to address portfolio selection is shown. Section 3 describes the methodology implemented to apply the MCHP and the Markowitz model for the selection of portfolios using stocks from the Mexican Stock Exchange. In Sect. 4, the results obtained by applying the methodology in the previous section are described. Finally, the main conclusions and future lines of research are discussed in Sect. 5.

2 Literature Review of Portfolio Selection

Portfolio selection is a very studied topic for practitioners and researchers. Some significant advances have been developed using approaches different from those classical financial methods. Aouni et al. [16] found 116 published studies of the portfolio selection problem that include MCDA methods, some focused on the evaluation or selection of assets to create a short list for investment and others focused on the construction and optimisation of investment portfolios.

This section aims to review the portfolio selection problem with Multicriteria Decision Making (MCDM) methods. The Tehran Stock Exchange (TSE) was investigated. Here specific criteria were selected, and a fuzzy analytic network process (FANP) was developed to assess and select portfolios on the TSE [17]. Another work [16] studies optimal equity portfolios with long-term horizons under heterogeneous risk aversion levels. It focuses on European stocks.

A method is introduced to select stocks based on financial ratio data, before forming the portfolio based on its financial performance through a multicriteria decision-making model using the Technique for Order of Preference by Similarity to Ideal Solution (TOPSIS) model. The optimal portfolio is generated using the Markowitz model [7]. Another article also shows the application of the Multiple Criteria Decision Aid (MCDA) method to deal with the multicriteria portfolio problem integrated into the Fuzzy AHP Fuzzy TOPSIS model [1].

Kao and Steuer [21] propose a methodology to calculate the value of the information, under the argument that in the mean variance model, the process to specify the expected returns of individual assets is based on the information, and it is decisive for the capital allocations to the different securities to be made correct. An illustrative application of the Taiwan Stock Exchange is presented. In [26], a combined possibility theory and a programming model by objectives is proposed that allows considering trade-offs between investors' preferences concerning several incommensurable objectives in an environment imprecisely.

This article [30] provides results in the area of the analytical derivation of the efficient set of a medium variance portfolio selection problem that has more than three criteria, beyond the mean and variance. Likewise, the problems addressed have two or more additional linear objectives that could be sustainability, dividend yield, liquidity and R&D as additional objectives.

Egyptian asset managers were surveyed to identify the necessary factors they consider to build a portfolio management model for investing in the Egyptian Stock Exchange (SEE) [15]. The survey concludes that the factors to consider are fair value, liquidity, sales growth, net profit growth, industry growth, bank interest rate, market capitalisation, the price per profit, dividends, cash dividend, volatility and Beta ratio portfolio market.

Hadhri and Ftiti [18] investigate whether it is profitable to invest in emerging markets. It focuses on the third classic moment used in asset allocation: realised bias. It uses an analysis based on asymmetry in the markets focused on the potential opportunities for diversification and the portfolio's investment strategy. In the work developed by [20], six online portfolio strategies' effectiveness is compared and applied to the Korean equity portfolio.

Three portfolio selection strategies for loss-averse investors are discussed [22]: semi-variance, conditional value at risk, and a combination of both risk measures. A novel version of the non-dominated sorting genetic algorithm II and of the strength Pareto evolutionary algorithms 2 to tackle this optimization problem. These algorithms' effectiveness is compared with two alternatives from the literature of five publicly available data sets. In [31] is analysed the calculating method and applicable conditions of the Markowitz risk remuneration model, Arrow–Pratt risk remuneration model and Jia and Dyer standard risk model. This paper gives the investment model optimised by mean entropy and portfolio model, including transaction cost.

3 Multiple Criteria Hierarchical Process and Markowitz Model

3.1 *Multiple Criteria Hierarchical Process*

The Multiple Criteria Hierarchical Process is based on MultiCriteria Analysis for Decision Making (MCDA). Corrente et al. [13] point out that it is frequent that in cases of practical application a hierarchical structure of criteria is presented; such is the case of the problem treated in this article, that given the description of the problem through the criteria of decision (financial indicators), the hierarchical nature of its structure is observed. Therefore, we consider appropriating the use of a new method, the Multiple Criteria Hierarchical Process (MCHP). Alvarez et al. [3] also use the MCHP method to analyse the competitiveness of the regions in Mexico since it is useful to generate a hierarchical structure that allows creating subgroups of a large number of criteria. In Alvarez et al. [4], the MCHP was applied to analyse the innovation capacity of Mexican regions.

In the MCDA process, the definition of a set of alternatives $A = \{a_1, a_2, ..., a_m\}$ and a coherent family of criteria $G = \{g_1, g_2, ..., g_n\}$ is carried out. Comprehensive modelling of preference is developed by any MCDA method as an aggregation procedure. The method generates a recommendation in a ranking format of alternatives in descending order from the best to the worst.

Corrente et al. [12] proposed a hierarchical version of the ELECTRE III method. The notation used in this process is explained as follows.

G is a comprehensive set of all criteria at all considered levels in the hierarchy.

G_0 is the root of the criterion.

l_G is the set of indices of the criteria in G.

$E_G \subseteq I_G$ is the set of indices of elementary criteria.

g_r is the generic non-root criterion (where r is a vector with length equal to the level of the criterion).

$g_{(r,1)}, \ldots, g_{(r,n(r))}$ are the immediate subcriteria of criterion g_r (located at the level below g_r).

$E(g_r)$ is the set of indices of all the elementary criteria descending from g_r.

$E(F)$ is the set of indices of the elementary criteria descending from at least one criterion in the subfamily $F \subseteq G$ (that is, $E(F) = U_{g_r \in F} E(g_r)$).

G_r^l is the set of subcriteria of g_r located at level l in the hierarchy (belonging to g_r).

L is the number of levels of the hierarchy $l = 1, \ldots, L$

3.2 The Hierarchical ELECTRE III Method

The ELECTRE method is carried out in two steps. In the first step, the aggregation of preference information is developed, constructing a decision model in the valued outranking relation. In the second step, the valued outranked relation is exploited by the distillation process, generating a complete or partial ranking of alternatives.

For each elementary criterion $g_t \in E(g_r)$.

The elementary concordance index, for each elementary criterion g_t, is

$$\varphi_t(a,b) = \begin{cases} 1 & if\ g_t(b) - g_t(a) \le q_t, (aS_tb) \\ \frac{p_t - [g_t(b) - g_t(a)]}{p_t - q_t} & if\ q_t < g_t(b) - g_t(a) < p_t, (bQ_ta) \\ 0 & if\ g_t(b) - g_t(a) \ge p_t, (bP_ta) \end{cases} \quad (1)$$

The discordant elementary index, for each elementary criterion g_t, is

$$d_t(a,b) = \begin{cases} 1 & if\ g_t(b) - g_t(a) \ge v_t \\ \frac{[g_t(b) - g_t(a)] - p_t}{v_t - p_t} & if\ p_t < g_t(b) - g_t(a) < v_t \\ 0 & if\ g_t(b) - g_t(a) \le p_t \end{cases} \quad (2)$$

The partial concordance index for each non-elementary criterion g_r, is

$$C_r(a,b) = \frac{\sum\limits_{t \in E(g_r)} w_t \varphi_t(a,b)}{\sum\limits_{t \in E(g_r)} w_t} \quad (3)$$

Partial credibility index is

$$\sigma_r(a,b) = \begin{cases} C(a,b) \times \prod_{g_t \in E(g_r)} \frac{1 - d_t(a,b)}{1 - C_r(a,b)} \; if \; d_t(a,b) > C_r(a,b) \\ C(a,b) \; otherwise \end{cases} \quad (4)$$

The distillation method is used as an exploitation procedure of the fuzzy outranking relation. For the pairs $a,\ b \in A$ in the hierarchical process, the alternatives are ranked in a partial or complete preorder on the non-elementary criterion g_r. For a further description of the distillation process, see [28].

The ELECTRE-III method is a well-known approach applied to a diversity of MCDM problems. In Alvarez et al. [2] are analysed municipal districts to construct a water supply system. In López et al. [25], a group of decision-makers evaluates the technological packages available to an agricultural company to select the most appropriate one.

3.3 The Markowitz Model

The Modern Portfolio Theory is based on Harry Markowitz's original work "Portfolio Selection" focusing on the main rule that the investor considers the expected return as desirable and the return variation as undesirable [27].

Portfolio Profitability is defined as the weighted sum of the individual returns of the selected assets of the portfolio, multiplied by the weight of assets in the portfolio [23].

To evaluate the performance of the portfolio R_p, Eq. (5) is applied:

$$R_p = \sum_{i=1}^{n} w_i R_i \quad (5)$$

where R_i is the performance of each asset, and w_i is the weight of R_i in the portfolio.

The portfolio risk can be expressed as the variance of the portfolio, which is calculated based on the variation of individual assets and the correlation between them. Portfolio risk is expressed in Eq. (6):

$$Var(P) = \sigma_p^2 = \sum_{i=1}^{n} w_i^2 \sigma_i^2 + 2 \sum_{i=1}^{n} \sum_{\substack{i=1 \\ j \neq i}}^{n} w_i w_j \sigma_{ij} \quad (6)$$

where σ_p^2 is the risk of the portfolio (variance), w_i is the percentage of the weight of asset i in the portfolio, σ_i^2 is the variance of the return of asset i, w_j is the percentage of the weight of asset j in the portfolio and $\sigma_{i,j}$ is the covariance between asset i and asset j.

Equation (7) minimises the risk subject to a given level of performance:

$$\begin{aligned} &Min\, \sigma_p^2 \\ &s.a.\ \mathrm{R}_p = R^* \end{aligned} \tag{7}$$

where R_p is the portfolio return R^*.

Given the returns of the considered assets, a portfolio return (R^*) must be established, which in turn will define the proportion of each share in the portfolio. The covariance is determined by Eq. (8):

$$COV(R_i, R_j) = \sigma_{ij} = \frac{1}{n}\sum_{i=1}^{n}[R_i - \mu_i][R_j - \mu_i] \tag{8}$$

Due to the difficulty in interpreting the magnitude of the covariance, correlation is used to measure the degree of joint movement between two variables or the linear relationship between both. It takes values in -1 and $+1$ and is determined by Eq. (9):

$$Corr(R_i, R_j) = p_{ij} = \frac{\sigma_{ij}}{\sigma_i \sigma_j} \tag{9}$$

where p_{ij} is the correlation between asset i and asset j, σ_{ij} is the covariance between asset i and asset j and σ_i and σ_j are the volatilities of i and asset j, respectively.

Pearson's correlation coefficient is calculated based on the observed returns, as expressed in Eq. (10):

$$Corr(x_i, y_j) = \frac{\sum_{i=1}^{n}(x_i - \mu_x)(y_i - \mu_y)}{\sqrt{\sum_{i=1}^{n}(x_i - \mu_x)^2 \sum_{i=1}^{n}(y_i - \mu_y)^2}} \tag{10}$$

A positive linear association coefficient, very close to 1, means that a large part of the simultaneous variability is being explained by the joint. If this were equal to zero, it would indicate that the covariance is zero or that one of the assets has no variability. When the sign of the correlation coefficient is negative, the absolute value indicates the percentage of the simultaneous variability that is explained by the joint variability. Therefore, the closer it is to -1, the greater the elimination of risk that is achieved with the combination of these assets in a portfolio.

The efficient frontier represents the optimal combinations of return and risk. The optimal portfolio combinations contain the highest return, given a level of risk or the lowest risk given an expected return. The graph presented in Fig. 1 shows which efficient combinations are inefficient (portfolios X, Y and Z), combinations that are not feasible (portfolio U) and combinations that are inefficient (portfolio T). So, starting at the lowest point of variation, a higher return on investment is only possible with higher risk, and the level of risk she is willing to accept is up to the investor.

Fig. 1 Efficient frontier

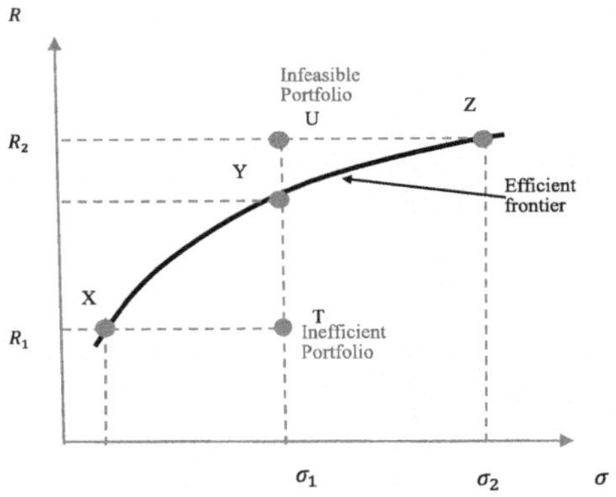

4 Multiple Criteria Hierarchical Process and Markowitz Model for Portfolio Selection in the Mexican Stock Exchange

4.1 Data Description

For the current research, the profitability ratios, asset turnover ratios, financial leverage ratios, liquidity ratios, market value ratios and risk ratios were used to evaluate the performance of the companies [29]. Both risk indicators [23] are considered, the Beta of the stock and volatility. These five groups of indicators are listed as follows: Profitability ratios, Asset turnover ratios, Financial ratios, Market value indicators and the Beta of the shares. For a more detailed description of the indicators, see [9].

4.2 Application MCHP and Markowitz Model

The first stage of the portfolio selection corresponds to the evaluation of the Mexican stocks. It is carried out with the Multiple Criteria Hierarchical Process (MCHP) explained in Sect. 3.2. The portfolio selection for Mexican Stock Exchange presents a structure of three levels in their financial indicator. The indicators allow the analysis of the companies' performance and their shares in the stock market. Figure 2 shows the scheme of the financial indicators' hierarchy. In Level 1, there are two macrocriteria, book performance and market. In Level 2, four subsets of criteria are defined for book performance: profitability, financial leverage, assets turnover and liquidity. Moreover, two subsets of criteria are defined for market macrocriterion: return and risk. Table 1 shows the complete list of macrocriteria, subsets of criteria and elementary criteria. For the portfolio selection problem, there are 22 elementary criteria evaluating Mexican stocks.

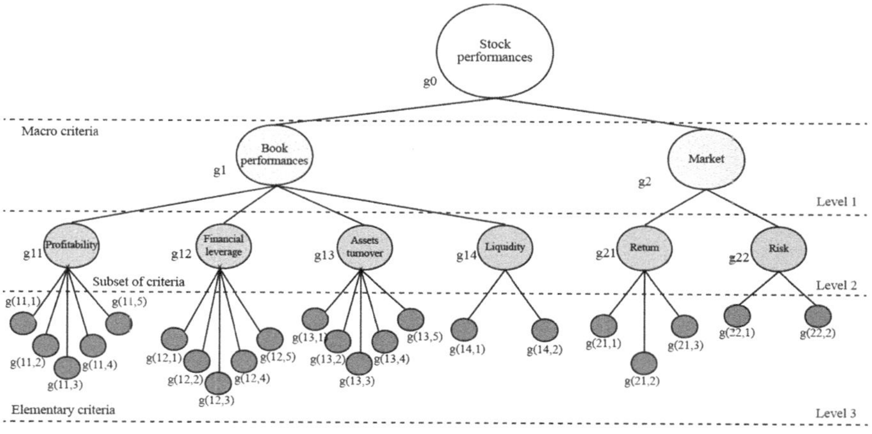

Fig. 2 The hierarchical structure of the stock performance

In addition, some parameters are defined for each elementary criterion for the hierarchical ELECTRE III method. The indifference threshold (q_t), the preference thresholds (p_t) and the weight (w_t) are defined for each elementary criterion (g_t). The defined parameters for the hierarchical ELECTRE III are listed in Table 2. For a broader description of the MCHP and ranking of stocks of each subset of criteria, see [9].

BMV includes 140 companies that are quoted on a stock exchange. Also, the BMV generates a Prices and Quotes Index (PQI) of 35 companies evaluated with better performance compared with the rest. It is worth mentioning that the BMV Index seeks to measure the performance of the larger and liquidity shares listed in the BMV and is designed to provide a representative, reversible and replicable measure of the Mexican stock market. The ranking of the 20 companies with the best performance is shown in Table 3. In comparison with the Mexican stock exchange index, the first 10 positions of the ranking in Table 3 include 80% of top companies from the index. The 20 ranked companies from the MCHP are used for the second stage of the portfolio selection described below.

4.3 *Markowitz Model Based on Multicriteria Ranking*

In the second stage, the Markowitz model is applied. The desired profitability is set to find the portfolio that produces this profitability with the least possible variance. Equal participation is considered for the 20 shares (5%).

The variance is minimised with the use of Solver with the following restrictions: The participations total 1, the average return of the portfolio is equal to 0.60% and the participations are not negative.

Table 1 Macrocriteria and elementary criteria for the stock selection problem

Index	Macro criteria	Index	Macro criteria	Index	Elementary criteria
1	Company performance	11	Profitability	(11, 1)	Return on Capital (ROE) (%)
				(11, 2)	Return on Sales (ROS)
				(11, 3)	Return on Assets (ROA) ($)
				(11, 4)	Economic Value Added (EVA) ($)
				(11, 5)	Dupont Identity (DI) (%)
		12	Financial leverage	(12, 1)	Debt (%)
				(12, 2)	Long-term Debt ((LTD) (%)
				(12, 3)	Debt to Capital (DTC) (%)
				(12, 4)	Multiplied of Capital (CMU) (%)
				(12, 5)	Interest coverage ratio (ICO) (Number of times)
		13	Assets turnover	(13, 1)	Turnover stocks (OTO) (Number of times)
				(13, 2)	Accounts receivable turnover (RAR) (number of times)
				(13, 3)	Asset Turnover (ATO) (Number of times)
				(13, 4)	Fixed asset turnover (TFA) (number of times)
				(13, 5)	Asset turnover frequency (FAT) (number of days)
		14	Liquidity	(14, 1)	Current Ratio (CRA) (number of times)
				(14, 2)	Quick ratio or acid test (QRA) (number of times)

(continued)

Table 1 (continued)

Index	Macro criteria	Index	Macro criteria	Index	Elementary criteria
2	Market	21	Return	(21, 1)	Price-Earnings (P / E) ($)
				(21, 2)	Market-Books (M / B) ($)
				(21, 3)	Earnings Per Share (EPS) ($)
		22	Risk	(22, 1)	Beta
				(22, 2)	Volatility (%)

Table 2 Parameters for the multicriteria method

No.	Label	Direction	wt	qt	pt
1	g(11, 1)	Max	0.024055	0.01	0.05
2	g(11, 2)	Max	0.019244	0.05	0.1
3	g(11, 3)	Max	0.014433	0.05	0.1
4	g(11, 4)	Max	0.028866	10,000	20,000
5	g(11, 5)	Max	0.009622	0.01	0.05
6	g(12, 1)	Min	0.015402	0.1	0.15
7	g(12, 2)	Min	0.007701	0.1	0.15
8	g(12, 3)	Min	0.012835	0.1	0.15
9	g(12, 4)	Max	0.005134	0.5	1
10	g(12, 5)	Max	0.010268	0.5	1
11	g(13, 1)	Max	0.007344	0.5	1
12	g(13, 2)	Max	0.011016	0.5	1
13	g(13, 3)	Max	0.014688	0.5	1
14	g(13, 4)	Max	0.01836	1	2
15	g(13, 5)	Min	0.022032	180	360
16	g(14, 1)	Max	0.064974	0.3	0.5
17	g(14, 2)	Max	0.054026	0.2	0.4
18	g(21, 1)	Max	0.07296696	3	5
19	g(21, 2)	Max	0.054945	0.5	1
20	g(21, 3)	Max	0.09186804	0.2	1
21	g(22, 1)	Min	0.24036012	0.002	0.005
22	g(22, 2)	Max	0.24036012	0.2	0.5

Table 3 Overall ranking (g) of the portfolio selection problem

Position	Company
1	*a26, a39
2	*a65
3	*a24
4	*a13
5	*a9
6	a32
7	a57
8	*a73
9	*a68
10	*a88
11	a96
12	*a67
13	*a61
14	*a63
15	a11
16	*a83
17	a56
18	a42
19	a116
20	a71

*Companies from the Prices and Quotes Index (PQI)

By applying this procedure, an efficient boundary point can be defined. This point represents a portfolio with the following results: Profitability = 0.6%, Variance = 0.00021798 and Portfolio Standard Deviation = 1.48%. Likewise, the portfolio that produces this result is composed of the participation of the shares that are presented in Table 4.

The process of identifying stock participation in the efficient portfolios, the minimum variance and, therefore, the standard deviation, with the desired profitability, allows determining points that make up the efficient frontier. Various portfolios were calculated establishing different levels of profitability, represented by the points on the efficient frontier (see Fig. 3 and Table 5).

The efficient frontier shows the combinations of assets resulting from applying the Markowitz model, by locating an efficient portfolio that will generate the desired return by setting a target risk level and whose return cannot be increased further without increasing risk. In this case, a border was obtained with 10 portfolios made up of the first 20 stocks in the ranking obtained in the first stage from the stocks listed on the Mexican Stock Exchange (BMV).

Table 5 presents 11 combinations, where the profitability, risk and weightings of these stocks can be observed in the composition of each portfolio. For example, with

Table 4 Resulting Stock Participation

No.	Action	Participation (%)
1	PINFRA	0.00
2	*CMOCTEZ	9.76
3	GRUMAB	0.00
4	OMAB	0.00
5	GAPB	0.00
6	ASURB	0.00
7	VESTA	0.00
8	BACHOCOB	0.00
9	WALMEX	0.00
10	*BOLSAA	2.98
11	*GPROFUT	20.68
12	KIMBERA	0.00
13	*CERAMICB	2.04
14	*MEGACPO	14.32
15	AC	0.00
16	*CYDSASA A	9.08
17	POSADASA	0.00
18	*MINSA	41.14
19	CIEB	0.00
20	LIVEPOLC-1	0.00
	Total	100

*Selected actions with participation > 0

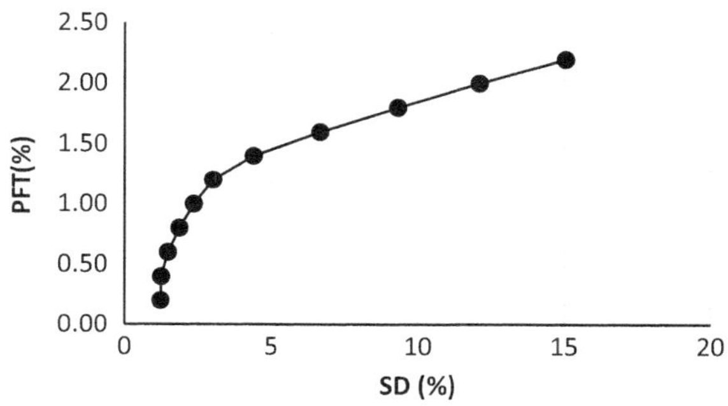

Fig. 3 Resulting efficient frontier

portfolio 1 a return of 0.20% is obtained at a risk level of 1.23%, represented by the standard deviation and with the following composition of shares: CMOCTEZ (7%), BACHOCOB (1%), GPROFUT (11%), KIMBERA (5%), MEGACPO (5%), CYDSASAA (2%) and MINSA (68%).

Table 5 Stock participation in the efficient portfolios

Portfolios	1	2	3	4	5	6	7	8	9	10	11
PFT (%)	0.20	0.40	0.60	0.80	1.00	1.20	1.40	1.60	1.80	2.00	2.20
SD (%)	1.23	1.25	1.48	1.87	2.36	3.01	4.39	6.67	9.32	12.12	15.05
Stocks participation in the portfolio (%)											
PINFRA	0	0	0	0	0	0	0	0	0	0	0
CMOCTEZ	7	11	10	7	7	0	0	0	0	0	0
GRUMAB	0	0	0	0	0	0	0	0	0	0	0
OMAB	0	0	0	0	0	0	0	0	0	0	0
GAPB	0	0	0	0	0	0	0	0	0	0	0
ASURB	0	0	0	0	0	0	0	0	0	0	0
VESTA	0	0	0	0	0	0	0	0	0	0	0
BACHOCOB	1	0	0	0	0	0	0	0	0	0	0
WALMEX	0	0	0	0	0	0	0	0	0	0	0
BOLSAA	0	0	3	6	9	10	13	7	1.4	0	0
GPROFUT	11	15	21	27	33	39	42	40	36.7	22	3
KIMBERA	5	0	0	0	0	0	0	0	0	0	0
CERAMICB	0	0	2	5	6	0	0	0	0	0	0
MEGACPO	5	9	14	19	24	19	0	0	0	0	0
AC	0	0	0	0	0	0	0	0	0	0	0
CYDSASAA	2	5	9	13	16	23	22	12	2.6	0	0
POSADASA	0	0	0	0	0	0	0	0	0	0	0
MINSA	68	59	41	24	5	0	0	0	0	0	0
CIEB	0	0	0	0	0	9	23	41	59.4	78	97
LIVEPOLC1	0	0	0	0	0	0	0	0	0	0	0

PFT = Profitability
ST = Standard deviation

5 Conclusions

The portfolio selection problem was treated in a comprehensive manner, in its two stages. In the first stage, the portfolio selection problem was addressed with the Multiple Criteria Hierarchical Process (MCHP). A hierarchical structure of multiple criteria of 3 levels of the problem was designed, and the data processing is described using the hierarchical method ELECTRE-III to generate the ranking of actions based on financial indicators. The ranking of shares of the Mexican Stock Exchange (BMV) generated by the MCHP made it possible to generate rankings in subgroups of criteria and a global ranking of the shares of the BMV.

In the second stage the Markowitz model was applied through the Solver tool, to obtain optimal portfolios and build an efficient frontier with the 20 stocks located in the first positions of the ranking gained in the first stage.

The research's main contribution is the integration of two different approaches to solve the portfolio selection problem. The hierarchical version of the ELECTRE-III from the MCDA approach and the classical Markowitz model from the financial approach were integrated. The former helps to incorporate in the aggregation model the investor preferences. The latter allows the analysis of the profitability of portfolios. The main findings of this integration are related to the efficient manner of evaluating assets regarding the complete set of criteria and analysing subgroups of criteria to know the asset's performance. The applied approaches show the portfolio selection is profitable, and the current integration methods can be implemented in other stock exchanges.

In line with this work, a future research field is visualised, and the inclusion of other indicators, in particular qualitative ones, could be tested for the evaluation of stock shares through the application of the fuzzy logic methodology, with linguistic labels [24]. The methodology used could be applied to other stock markets considering different performance criteria of the company and the market and thus prove its efficiency in those markets. On the other hand, the Markowitz model's portfolio selection problem can be improved, optimising stock participation [5, 14]. It is a promising field of research to find a better selection of investment in each portfolio stock.

References

1. Aldalou E, Perçin S (2018) Financial Performance Evaluation of Turkish Airline Companies Using Integrated Fuzzy Ahp Fuzzy Topsis Model*. Uluslararası İktisadi ve İdari İncelemeler Dergisi 18(EYI Special Issue):583–598. https://doi.org/10.18092/ulikidince.347925
2. Alvarez PA, Morais DC, Leyva López JC, de Almeida AT (2022) An ELECTRE III based consensus-reaching process to improve a collective solution. Int Trans Oper Res 29(2):1048–1088. https://doi.org/10.1111/itor.12798
3. Alvarez P, Muñoz-Palma M, Miranda-Espinoza L, Lopez-Parra P, León-Castro E (2020) Enfoque multicriterio jerárquico para el análisis de la competitividad de las regiones en México. Inquietud Empres 20(2):29–51. https://doi.org/10.19053/01211048.11408
4. Alvarez PA, Valdez C, Dutta B (2022) Analysis of the innova-tion capacity of mexican regions with the multiple criteria hierarchy process. Socio-Econ Plan Sci 84. https://doi.org/10.1016/j.seps.2022.101418
5. Amiri M, Ekhtiari M, Yazdani M (2011) Nadir compromise programming: a model for optimization of multi-objective portfolio problem. Expert Syst Appl 38(6):7222–7226
6. Aouni B et al (2018) On the increasing importance of multiple criteria decision aid methods for portfolio selection. J Oper Res Soc 69(10):1525–1542. Taylor & Francis. https://doi.org/10.1080/01605682.2018.1475118
7. Aqilah N, Fauzi M, Ismail M (2019) Applicability of TOPSIS Model and Markowitz Model. J Phys: Conf Ser 1212(012032). https://doi.org/10.1088/1742-6596/1212/1/012032
8. Basilio M et al (2018) Investment portfolio formation via multicriteria decision aid: a Brazilian stock market study. J Model Manag 13(12):394–417. https://doi.org/10.1108/JM2-02-2017-0021
9. Bernal MR, Alvarez PA, Muñoz M, Leon-castro EA, G-CDA (2020) Multicriteria hierarchical aproach for portfolio selection in stock exchange. J Intell Fuzzy Syst 40:1945–1955

10. Boonjing V, Boongasame L (2016) Combinatorial Portfolio selection with the ELECTRE III method: case study of the stock exchange of Thailand (SET). Proc 2016 Fed Conf Comput Sci Inf Syst 8(4):719–724. https://doi.org/10.15439/2016f228
11. Bouri A, Martel JM, Chabchoub H (2003) A multi-criterion approach for selecting attractive portfolio. J Multicriteria Decis Anal II(February):269–277. https://doi.org/10.1002/mcda.334
12. Corrente S et al (2017) A robust ranking method extending ELECTRE III to hierarchy of interacting criteria, imprecise weights and stochastic analysis. Omega (United Kingdom) 73:1–17. https://doi.org/10.1016/j.omega.2016.11.008
13. Corrente S, Greco S, Słowiński R (2012) Multiple criteria hierarchy process in robust ordinal regression. Decis Support Syst 53(3):660–674. https://doi.org/10.1016/j.dss.2012.03.004
14. Ehrgott M, Klamroth K, Schwehm C (2004) An MCDM approach to portfolio optimization. Eur J Oper Res 155:752–770. https://doi.org/10.1016/S0377-2217(02)00881-0
15. Elselmy HS, Ghoneim A, Elkhodary IA (2019) Portfolio selection factors: Egypt equity market case study. In: ACM international conference proceeding series, pp 212–216. https://doi.org/10.1145/3328833.3328858
16. Erdemlioglu D, Joliet R (2019) Long-term asset allocation, risk tolerance and market sentiment. J Int Financ Mark, InstS Money 62:1–19. Elsevier B.V. https://doi.org/10.1016/j.intfin.2019.04.004
17. Galankashi MR, Rafiei FM, Ghezelbash M (2020) 'Portfolio selection : a fuzzy-ANP approach', Financial Innovation. Financ Innov 6(17)
18. Hadhri S, Ftiti Z (2019) Asset allocation and investment opportunities in emerging stock markets: Evidence from return asymmetry-based analysis. J Int Money Financ 93:187–200. Elsevier Ltd. https://doi.org/10.1016/j.jimonfin.2019.01.002
19. Hurson C, Zopounidis C (1997) On the use of multicriteria decision aid methods to portfolio selection. In: J. Clímaco (eds) Multicriteria analysis. Springer, Berlin, pp 496–507. https://doi.org/10.1007/978-3-642-60667-0_47
20. Jeong T, Kim K (2019) Effectiveness of F-SCORE on the loser following online portfolio strategy in the korean value stocks portfolio. Am J Theor Appl Bus 5(1):1–13. https://doi.org/10.11648/j.ajtab.20190501.1
21. Kao C, Steuer RE (2016) Value of information in portfolio selection, with a Taiwan stock market application illustration. Eur J Oper Res 253(2):418–427. Elsevier B.V. https://doi.org/10.1016/j.ejor.2016.02.011
22. Kaucic M (2019) 'Portfolio optimization by improved NSGA-II and SPEA 2 based on different risk measures', Financial Innovation. Financ Innov 5:5–26. https://doi.org/10.1186/s40854-019-0140-6
23. De Lara A (2009) Medición y Contro de Riesgos Financieros. 3a. ed. Limusa
24. Liern V, Pérez-Gladish B (2018) 'Ranking corporate sustainability: a flexible multidimensional approach based on linguistic variables'. Int Trans Oper Res 25(3):1081–1100
25. López JCL, Carrillo PAÁ, Valenzuela OA (2018) A multicriteria group decision model for ranking technology packages in agriculture. In: En Cruz Corona C (eds) Soft computing for the sustainability science. Springer. https://doi.org/10.1007/978-3-319-62359-7_7
26. Mansour N, Cherif MS, Abdelfattah W (2019) Multi-objective imprecise programming for financial portfolio selection with fuzzy returns. Elsevier Ltd., Expert systems with applications. https://doi.org/10.1016/j.eswa.2019.07.027
27. Markowitz H (1952) 'Portfolio Selection', The theory and practice of investment management: asset allocation, valuation, portfolio construction, and strategies, 2nd edn, vol 7(1), pp 77–91. https://doi.org/10.1002/9781118267028.ch3
28. Marzouk MM (2011) ELECTRE III model for value engineering applications. Autom Constr 20(5):596–600. https://doi.org/10.1016/j.autcon.2010.11.026
29. Ortega A (2008) Planeación financiera estratégica. Mc Graw Hill
30. Qi Y, Steuer RE (2018) On the analytical derivation of efficient sets in quad-and-higher criterion portfolio selection. Ann Oper Res. 293(2):521–538. Springer US. https://doi.org/10.1007/s10479-018-3101-y

31. Zhu J (2019) Theoretical and empirical study on risk measurement method statistics and portfolio model theoretical and empirical study on risk measurement method statistics and portfolio model. J Phys: Conf Ser 1213:042009. https://doi.org/10.1088/1742-6596/1213/4/042009
32. Zopounidis C, Despotis DK, Kamaratou I (1998) Portfolio selection using the ADELAIS multiobjective linear programming system. Comput Econ II:189–204
33. Zopounidis C, Doumpos M (2002) Multi-criteria decision aid in financial decision making : methodologies and literature review. J Multicriteria Decis. Anal. II(December):167–186

Chapter 12
Forecasting the Exchange Rate in Diffuse Time Series by Hybridization of the OWA Operator in an Adaptive ARANFIS Neuronal Network

Diego Quintero-Avellaneda, Fabio Blanco-Mesa, Ernesto León-Castro, Andrés Alejandro Fernández Rosas, and Dianny Fernandez-Samacaa

1 Introduction

Traditionally, investment capital negotiators carry out their decision-making based on analyses of stock market data, political decisions, and other external variables that may affect the value of their assets. A multitude of variables are involved in the decision-making process [2, 22, 24], and the knowledge, experience, and expectations of the decision-makers are key factors which can influence risk forecasting to predict future conditions [5, 10, 38]; however, the ability of some tools used for capital investment to incorporate these factors is limited. To overcome this, Yager [42] recommends the use of the ordered weighted average (OWA), which can obtain optimal weights for the inferential attributes of each decision-maker, that is, it allows us to underestimate or overestimate a problem based on the values assigned by the results of the decision-maker assessments [4, 20, 27].

On the other hand, by recognizing that the volatility in the stock market is not linear, the literature suggests recent approaches from the field of artificial intelligence through the adaptive neuro-fuzzy inference system (ANFIS), proposed by [18] as a forecasting method that recognizes that the relationship between the variables in the real-time series in many circumstances cannot be determined a priori based on their behavior. Furthermore, there are numerous studies on the hybridization of linear and nonlinear models. For example, [45] proposed a forecast time series method using a hybrid ARIMA and a neural network model, and [34] suggested a hybrid

D. Quintero-Avellaneda · F. Blanco-Mesa · A. A. F. Rosas · D. Fernandez-Samacaa
Escuela de Administración de Empresas, Facultad de Ciencias Económicas y Administrativas, Universidad Pedagógica y Tecnológica de Colombia; Av. Central del Norte, Tunja, Colombia

E. León-Castro (✉)
Unidad Navojoa, Instituto Tecnologico de Sonora, amon Corona sin numero Colonia ITSON, 85860 Navojoa, Sonora, México
e-mail: ernesto.leon243151@potros.itson.edu.mx

M. Sahni et al. (eds.), *Novel Developments in Futuristic AI-based Technologies*, Algorithms for Intelligent Systems, https://doi.org/10.1007/978-981-99-3076-0_12

ARIMA-ANFIS method using a two-stage approach. In addition, [6] proposed an autoregressive neuro-diffuse adaptation inference, a hybridized system of AR-ANFIS in which they implemented a one-stage approach. Finally, [8] postulate that the OWA operator can be used to reduce the dimension of file set data and employ a linear equation to merge the last periods of price values with the weights to make an aggregated input for the ANFIS model and produce a new forecast.

Therefore, this article presents the hybridization proposal that allows unifying the OWA operator with an AR-ANFIS system, recognizing the benefits of the OWA operator by generating different possible universes of action, which oscillate between the probability of the decision based on a trend in historical data and the probability of carrying out the decision-making process by soft elements, expressed as an underestimation of the history of the databases. Furthermore, the autoregressive characteristic (AR) of time series in financial prediction models is a key factor for understanding the existing probabilistic analytical techniques, which assess a high dimension of datasets to generate future values. In addition, based on the dimension of the information available, the systems that are best adapted for data processing are the ANFIS. Thus, it is beneficial to combine the transversal characteristics of the OWA operator with the conditional language (IF–THEN) of the ANFIS. In this way, a hybridized system of autoregressive neuro-diffuse adaptation inference with the OWA operator (AR-ANFIS ARANFIS-OWA) is obtained.

This study is structured as follows. In Sect. 2, the theoretical framework is presented; Sect. 3 presents the proposed method, with a step-by-step presentation of the considerations of the hybrid model. In Sect. 3, the results of the application of the proposed model are presented and compared with other traditional models. Finally, the main conclusions are presented in Sect. 4.

2 Theoretical Framework

Exchange rate forecasting has been an important issue for businesses and countries alike [26, 30]. The main problem with the actual methods is that the models based on the economic fundamentals are not that important short-term predictions [11, 25] because of which new models based on new machine learning methods are important to improve the error in the forecasting in these specific cases [37, 46].

The idea of using machine learning methods with exchange rate has been done before such as the cases of [33] that uses support vector machines (SVM) and neural networks (NW), [1] use sequential ridge regression and the exponentially weighted average strategy, [40] presents a Bagging-based on long short-term memory (LSTM-B), [35] compares the results obtained by SVM, NW, and Random Forest (RT) finding that RF is better than the other two techniques, and many more examples of machine learning can be found in the literature.

In the case of time series forecasting, the ARANFIS method has been used in different cases such as stock indexes [39], e-commerce [23], energy consumption

[17], water supply [21], and resource allocation [32]. As can be seen from the examples, the use of ARANFIS in time series has been applied to different subjects; in this paper, a hybridization of the OWA operator in an adaptive ARANFIS neuronal network is developed to improve the short-term forecasting of exchange rate.

3 Methodology

The proposed method to generate the hybridization of the OWA operator in an AR-ANFIS adaptive neural network is presented in this section. A hybridized system of autoregressive neuro-diffuse adaptation inference with the OWA operator (AR-ANFIS ARANFIS-OWA) is obtained. The method allows dealing with the complexity of the nonlinear financial time series and integrating optimism degree in forecasting exchange rate value problem, applied in the USD/COP currency price forecast. In this sense, this method considers overestimating or underestimating the problem.

First, volatility is considered to show the frequency and intensity of changes in the price of an asset, which allows for a certain interpretation of the state of the market, through the implementation of different models (i.e., with uncertainty and with a certain character of arbitrariness), to inform the decision-maker if the market is rising or falling. The problem is that, regardless of the arbitrariness, uncertainty exists in the question. Therefore, there is no guarantee that the probabilistic results are correct; they are the ones with the highest probability of occurrence, but the accuracy of the prediction is not verified. Hence, statistical models cannot always adequately fit stock datasets [19]. Thus, there may be many cases where the probabilistic results are not entirely satisfactory.

Second, it is useful to have the flexibility to account for whether the analyst wishes to arbitrarily overestimate or underestimate the results obtained from an analysis of the volatility because different types of decision-makers may have different outlooks: optimistic or pessimistic about probabilistic results. This is usually addressed by the OWA operator. Indeed, the OWA operator combines with a hybrid approach allowing to improve its results to incorporate knowledge of fuzzy time series data processing systems, such as ANFIS. On the one hand, knowledge based on subjectivity and attitude is considered. On the other hand, historical data is used as objective information.

Third, it is noteworthy that the high dimensionality of the data has a great deal of relevance. When the dataset are financial assets like stocks, they contain a large different variables, which depends on the indicators and techniques implemented for their reporting. In this sense, it is essential to merge the variables to avoid losing information from the original variables, which is important in forecasting models. Another consideration is that for multivariate prediction models, the computational complexity will increase much faster than the growth of the data dimension.

Based on the above concerns, the use of the OWA operator allows to aggregate the input information of the last price periods and the weight vectors given in the ANFIS

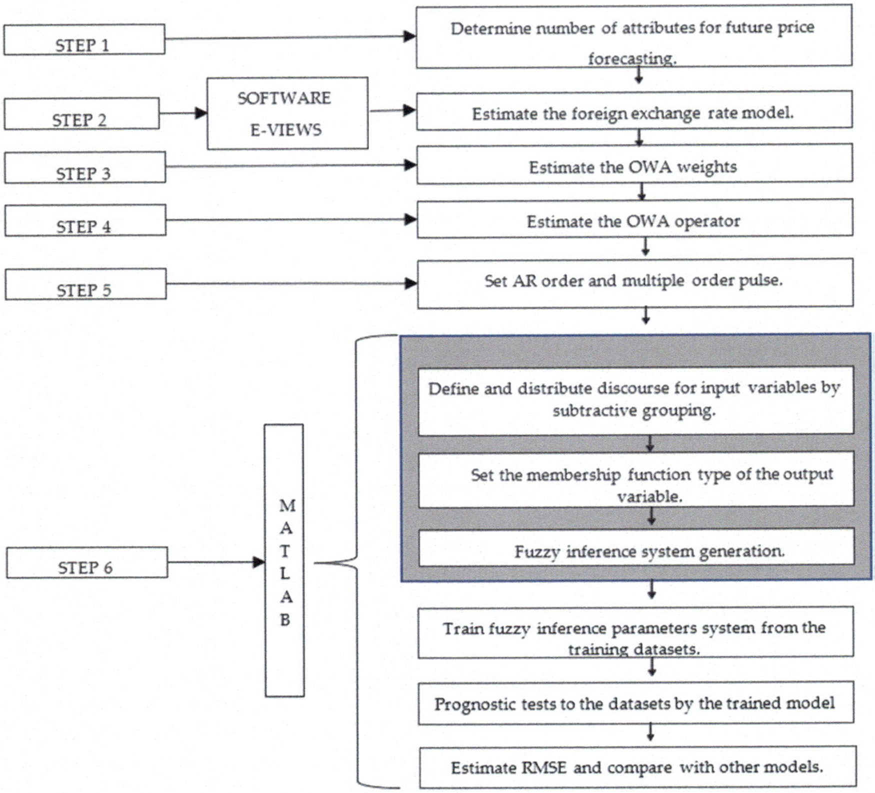

Fig. 1 Flow diagram to generate the hybridization of the OWA operator in an ARANFIS adaptive neural network

model. The OWA operator reduces data dimension in a linear equation using a simple method [15] argue that weighting-based technologies are more efficient methods for optimizing the time series model performance. In fact, the OWA operator allows to make a merging process of the information where each attribute is multiplied by a weight vector obtaining a single aggregated value. Hence, the linear computation of the OWA operator can reduce the dimension of the data. Considering these details, the process of the proposed method is presented (see Fig. 1).

4 Results

Findings show that the application of each proposed process is described. The analysis jointly applies the model based on the USD/COP exchange rate data to approach a forecast that includes not only the historical data, not necessarily assuming the same aversion to risk, thus incorporating not only the apparent states of the market but also the different decision models of each decision-maker.

Step 1. Determine the number of attributes for future price forecast.

In this stage, the prices related to the USD/COP exchange rate in the period from 2016 to 2018 were chosen for the proposed model, where a series of subsets of volatility data are used to train the system, which includes the values from January to October of the transaction days of each year, and the outstanding dataset, with data from November to December, is used to validate the test.

Period selection in the dataset is based on the research carried out by [7]. Thus, the periods of stock prices are $P(t-2)$, $P(t-1)$, and $P(t)$ to forecast $P(t+1)$. In a practical stock market, there is an argument that the influence degree of the past stock price would decrease, in this sense.

Step 2. Estimate the USD/COP model

$$P(t) > P(t-1) > P(t-2) \tag{1}$$

In the estimation of the USD/COP model, E-Views software is used to estimate the time series for different orders over 3 years. The least squares method is used in the construction of the model. The selected regression variables ((closing price (t–1) to closing price (t–3)) allow testing for each year (Table 1).

Step 3. Estimate the OWA weights.

Table 1 AR order and timing test (delay/period test)

Dependent Variable: CLOSE_PRICE				
Method: Least Squares				
Sample (adjusted): 2.638				
Included observations: 98 after adjustments				
Variable	Coefficient	Std. Error	t-Statistic	Prob
C	117.4171	84.00789	1.398082	0.1602
CLOSE_PRICE (−1)	1.004995	0.069541	14.97541	0.0000
CLOSE_PRICE (−2)	0.0035896	0.096345	0.040857	0.6764
CLOSE_PRICE (−3)	−0.063784	0.097432	−0.66934	0.5523
R-SQUARED	0.969834	Mean dependent var	3006.073	
Adjusted R-squared	0. 969,276	S.D. dependent var	734.0923	
S.E. of regression	119.9861	Akaike info criterion	10.93451	
Sum squared residual	3,255,721	Schwarz criterion	10.87569	
Log likelihood	−1348.958	F-statistic	2341.486	
Durbin-Watson stat	1.98024	Prob (F-statistic)	0.000000	

Source Own elaboration

In this stage, an adaptation made to Yager's OWA is presented in a polynomial equation, where [15] used Lagrange multipliers. Additionally, the associated weighting vector is obtained as follows:

$$\ln w_j = \frac{j-1}{n-1} \ln w_n + \frac{n-j}{n-1} \ln w_1 \Rightarrow w_j$$
$$= \sqrt[n-1]{w_1^{n-j} w_n^{j-1}}, \tag{2}$$

and

$$w_1[(n-1)\alpha + 1 - nw_1]^n = [(n-1)\alpha]^{n-1}[((n-1)\alpha - n)w_1 + 1]. \tag{3}$$

If

$$w_1 = w_2 = \ldots = w_n = \frac{1}{n} disp(w) = \ln n, (\alpha = 0.5) \tag{4}$$

then

$$w_n = \frac{((n-1)\alpha - n)w_1 + 1}{(n-1)\alpha + 1 - nw_1}, \tag{5}$$

where w_i is the weighting vector, n is the number of attributes, and α is the situation parameter. [7] perform the following procedure with (3) and (4) to obtain (5):

$$OWA(n, \alpha),$$

where n is the number of attributes and α is the situation parameter.
If $\alpha < 0.5$
then $\alpha = 1 - \alpha$
If $\alpha > 0.5$,
then

$$w_1[(n-1)\alpha + 1 - nw_1]^n = [(n-1)\alpha]^{n-1}[((n-1)\alpha - n)w_1 + 1]. \tag{6}$$

Then, the calculation in (5) is performed where w_n presents i = 2 with n–1; then, you obtain

$$w_i = \sqrt[n-1]{w_1^{n-i} w_n^{i-1}}. \tag{7}$$

Thus, the OWA operator is implemented and established on various α values, which represent an influence dataset considering different conditions in the stock market. Three weights are used: a) w1 (maximum value), w2 (average value), and

Table 2 OWA weights n = 3

Weights	⚜= 0.5	⚜ = 0.7	⚜ = 0.9
w_1	0.333	0.5540	0.8263
w_2	0.333	0.2920	0.1470
w_3	0.333	0.1540	0.0263

w3 (minimum value) between 0.5 and 1.0 to produce different degrees of influence on stock prices. The values are presented in Table 2.

Step 4. Estimate the OWA operator.

With the parameters of the degree of influence and share price, the linear equation is estimated using the subtractive clustering developed by [9], where a set T of *n* data points in a D-dimensional hyperspace is considered, where each data point w_i $(i = 1, 2 \ldots, n)$. $w_i = (x_i, y_i)$, where x_i denotes *p* input variables and y_i is the output variable. The potential value *Pi* of the data point is calculated using Eq. (8):

$$P_i = \sum_{j=1}^{n} e^{-\alpha \|w_i - w_j\|^2}, \tag{8}$$

where $\alpha = \frac{4}{r^2}$, *r* is the radius that defines a neighborhood of w_i, and ||…|| denotes a Euclidean distance.

A data point with a large number of data points around it would have a high potential value, and this is selected as the first cluster center.

Other cluster centers are generated by the potential P_i for each data point W_i in (9):

$$p_i = p_i - p_1^* \exp\left(-\beta \left\| w_i - w_1^* \right\|\right)^2, \tag{9}$$

where β is a positive constant, defining the neighborhood that will have measurable reductions in potential, w_1^* is the first cluster center, and p_1^* is its potential value, where [8] denotes the following:

$$A(t) = w_1 \times P(t) + w_2 \times P(t-1) + w_3 \times P(t-2), \tag{10}$$

where w_1, w_2, *and* w_3 denote diverse influence degrees for the three recent stock price periods, and $P(t)$, $P(t{-}1)$, $P(t{-}2)$, and $A(t)$ represent the aggregate value of the last three periods of stock prices with its related influence degrees.

This added value will be used in Step 6 as the first substep to generate the ANFIS forecast model.

Step 5. Set AR order and multiple order momentum.

With the results of the USD/COP estimation model, we obtain the reference data to determine the AR order. To set the one- and two-order impulses in predicting stock prices, the proposal of [41] is considered. To calculate the timing of multiple orders, two volatility variables are defined *Mo (t)* and *Mo (t – 1)* (see Table 3), and estimated by (11) and (12):

$$Mo\,(t) = USDCOP\,(t) - USDCOP\,(t - 1) \tag{11}$$

$$Mo(t - 1) = USDCOP(t - 1) - USDCOP(t - 2), \tag{12}$$

where *Mo (t)* is the impulse between time *t* and *–1*, and *Mo (t – 1)* is the impulse between time *t – 1* and *t – 2*.

Step 6. ANFIS forecasting model generation.

Here, forecast rules are created using the ANFIS model and the subtractive grouping method to obtain the final forecast on the future price. To do this, the following substeps must be carried out.

Step 6.1. Using subtractive clustering for the definition and discourse universe distribution for the inputs: Here, discourse universe for the input variables A (t) is defined and three recent periods of stock prices according to their minimum and maximum values are aggregated. After that, the discourse universe is divided to obtained linguistic intervals using subtractive grouping [9].

Table 3 Impulse of one and two orders of USD/COP

Date	USD/COP (T + 1)	Mo (T)	Mo (T–1)
04/01/2016	$ 3.149,47		
05/01/2016	$ 3.213,24		
06/01/2016	$ 3.203,86	63,77	
07/01/2016	$ 3.250,69	–9,38	63,77
08/01/2016	$ 3.287,28	46,83	–9,38
12/01/2016	$ 3.268,37	36,59	46,83
13/01/2016	$ 3.246,51	–18,91	36,59
14/01/2016	$ 3.235,45	–21,86	–18,91
15/01/2016	$ 3.240,71	–11,06	–21,86
19/01/2016	$ 3.293,94	5,26	–11,06
20/01/2016	$ 3.297,46	53,23	5,26
21/01/2016	$ 3.357,67	3,52	53,23
22/01/2016	$ 3.368,49	60,21	3,52
25/01/2016	$ 3.281,74	10,82	60,21
26/01/2016	$ 3.362,38	–86,75	10,82
27/01/2016	$ 3.375,80	80,64	–86,75
28/01/2016	$ 3.366,63	13,42	80,64
29/01/2016	$ 3.302,92	–9,17	13,42

Step 6.2. Set the membership function type of the output variable.: Here, membership functions of the linear type for the output variable are set. For example, there are two inputs (USDCOP (t), Mo (t)) and three linguistic intervals that are divided by subtractive grouping in each input variable. Thus, a Sugeno fuzzy model rule is considered below.

If $A_i = x$ *(USDCOP (t))*, and $B_i = y$ *(Mo (t))*, *then* $f_i = p_i\,x + q_i\,y + r_i$, where *x (USDCOP (t))*, y *(Mo (t))*, are linguistic variables, A_i y B_i, are the linguistic values (high, medium, low), and f_i denotes the *i*th. The output value, p_i, q_i, r_i, are the parameters *(i = 1,2,3)*.

Step 6.3. Fuzzy inference system generation.

With the linguistic results obtained on Step 6.2 the incoming and outgoing membership functions are obtained, the fuzzy rules are generated "if–then" valued by A_i, B_i and C_i that represent the membership functions which are used as "if-condition". Note that the membership function (f_i) is the "then" component.

Step 7. Parameters for the training datasets and forecasting.

Two different methods are used: the least squares method and backward propagation gradient descent method, with epoch at 50, simulation of 50 trials, or where the training error converges. The model will be determined when the stopping criterion is got and the forecasting can be seen in Fig. 3.

Step 8. Error estimation of the forecasting.

Here, RMSE values are estimated in the datasets tested using Eq. (13). Afterward, as an evaluation criterion for comparison with the other models the RMSE is used:

$$RMSE = \frac{\sqrt{\sum_i^n |real(t) - forecast(t)|^2}}{n}, \tag{13}$$

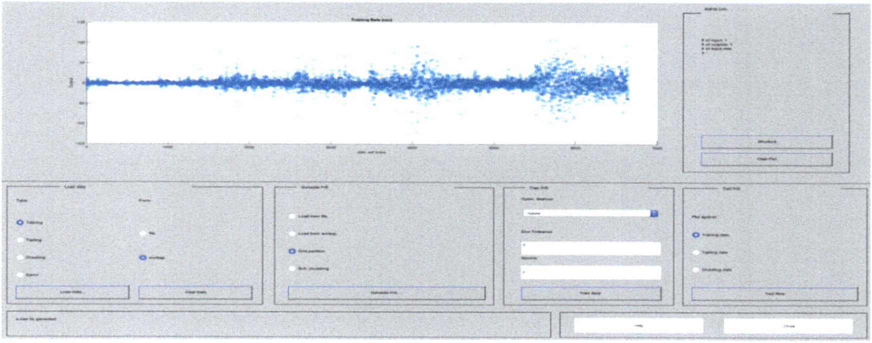

Fig. 2 ANFIS input image for USDCOP

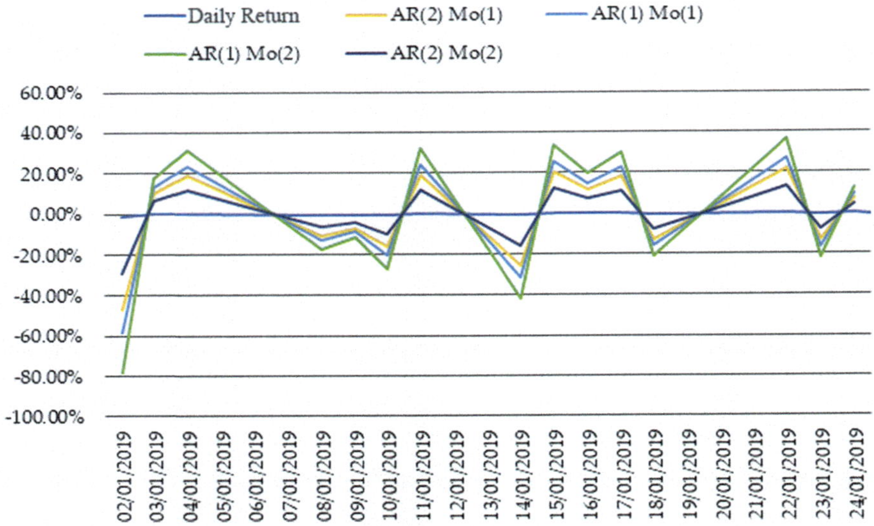

Fig. 3 Forecast USDCOP (t + 1)

where *real(t)* denotes the actual USDCOP value, *forecast(t)* denotes the predicted USDCOP value, and *n* is the amount of data. The comparison of the different results obtained from the RMSE is shown in Fig. 4.

The models described above are implemented in MATLAB, using some commands from the fuzzy logic inference system toolbox and the training functions

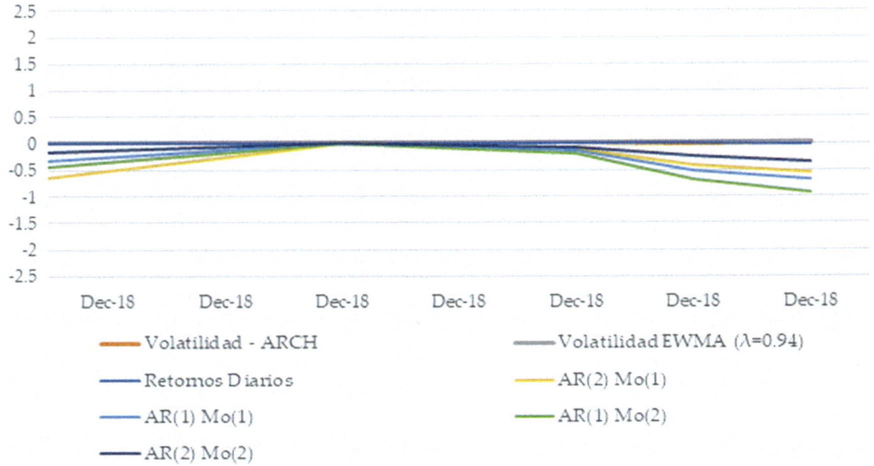

Fig. 4 Comparison of conventional models with the proposed model

for the ANFIS and Takagi–Sugeno systems. On the other hand, for the application of the least squares method, E-Views software was used.

The results in Fig. 4 show that the suggested model has a better performance in relation to RMSE than the ARCH models, demonstrating the advantage of the OWA operator. This can be attributed to the fact that autoregressive models do not capture the nonlinear characteristics typical of financial time series. Therefore, the proposed hybrid model can be used with greater reliability by capital investors and decision-makers to plan their investment decisions. Although the model is limited by its current parameters, it can be extended in the future to new parameterized families that include other variables that also influence the decision process.

5 Conclusions

In this research, we proposed the hybridization of two methods, the first the ARANFIS system that comes from the field of artificial intelligence, and the second the OWA operator, obtaining a hybridized system of autoregressive neuro-fuzzy adaptation inference with OWA operator (AR-ANFIS ARANFIS-OWA). The OWA operator is an important tool for dealing with multi-attribute data, showing differences from classical weighted methods. Since the OWA operator can adapt the weight of the characteristics in accordance with the situation of the decision-maker, as explained in Table 2, by adding different attribute values, it can generate a single added value that is used to generate a series of fuzzy if–then rules (see Table 3) from the ARANFIS forecasting model to forecast performance.

The results obtained show that the hybridized model surpasses compared methods for forecast accuracy, contributing to the applications of the OWA operator in the finance field similar to those proposed by [12, 13, 29]. This does not mean that autoregressive models such as ARIMA or GARCH are obsolete, but rather that, due to their structure, they cannot capture the nonlinear characteristics present in financial time series, such as stock index series, making it necessary to combine these methods with others that can model these characteristics and improve predictions.

Additionally, findings are useful and feasible for equity investors, decision-makers, and future research. Investors can use this "if–then" rule generated by ANFIS to discover the highest profit investment objective in the stock market. Furthermore, the most reasonable and comprehensible rules constructed by ANFIS can inform the qualitative understanding and knowledge. Hence, it is necessary to continue proposing and adapting prediction methods to the specific needs in different contexts, for example, the characteristics of stock indices, to achieve more accurate predictions, minimizing investor risk. One avenue is framing more complex theories that involve modern computational systems such as neural networks, fuzzy logic, neural fuzzy systems, genetic and evolutionary programming and chaos theory, and tools framed in the paradigm of artificial intelligence to different applications.

In future research, the model can be extended to new parameterized families that include expert and group decision models using prioritized OWA operator [14, 31,

44] or extend using a heavy OWA operator focusing on weighting vector [22, 28, 43]. Additionally, the use of new decision parameters that expand the optimization in terms of the variables that affect decision-making can be implemented [3, 16, 36].

Financial Statement This research received financial support from (a) Universidad Pedagogica y Tecnologica de Colombia, subvention number SGI 3134 and (b) FONDECYT Initation grant number 11190056.

Acknowledgements We thank the Intelligent and Expert Systems, Iberoamerican Computational Models. Thematic Network Number 522RT0130 in Iberoamerican Science and Technology for Development Program (CYTED) 2022–2025.

References

1. Amat C, Michalski T, Stoltz G (2018) Fundamentals and exchange rate forecastability with simple machine learning methods. J Int Money Financ 88:1–24
2. Bernal M, Anselmo Alvarez P, Muñoz M, Leon-Castro E, Gastelum-Chavira DA (2021) A multicriteria hierarchical approach for portfolio selection in a stock exchange. J Intell Fuzzy Syst 40(2):1945–1955
3. Better M, Glover F, Kochenberger G, Wang H (2008) Simulation optimization: applications in risk management. Int J Inf Technol Decis Mak 7(04):571–587
4. Blanco-Mesa F, Gil-Lafuente AM, Merigó JM (2018)New aggregation operators for decision-making under uncertainty: an applications in selection of entrepreneurial opportunities. Technol Econ Dev Econ 24(2):335–357
5. Blanco-Mesa F, Merigó JM, Gil-Lafuente AM (2017) Fuzzy decision making: a bibliometric-based review. J Intell & Fuzzy Syst 32(3):2033–2050
6. Chang JR, Wei LY, Cheng CH (2011) A hybrid ANFIS model based on AR and volatility for TAIEX forecasting. Appl Soft Comput 11(1):1388–1395
7. Chen T-L, Cheng C-H, Teoh H-J (2008) High-order fuzzy time-series based on multi-period adaptation model for forecasting stock markets. Phys A: Stat Mech Its Appl 387(4):876–888
8. Cheng C-H, Wei L-Y, Liu J-W, Chen T-L (2013) OWA-based ANFIS model for TAIEX forecasting. Econ Model 30:442–448
9. Chiu SL (1994) Fuzzy model identification based on cluster estimation. J Intell Fuzzy Syst 2(3):267–278
10. Dingli A, Fournier KS (2017) Financial time series forecasting-a machine learning approach. Mach Learn Appl: Int J 4(1/2):3
11. Engel C, West KD (2005) Exchange rates and fundamentals. J Polit Econ 113(3):485–517
12. Engemann KJ, Miller HE, Yager RR (1996) Decision making with belief structures: an application in risk management. Internat J Uncertain Fuzziness Knowl-Based Syst 4(01):1–25
13. Espinoza-Audelo L, Aviles-Ochoa E, Leon-Castro E, Blanco-Mesa F (2019) Forecasting performance of exchange rate models with heavy moving average operators. Fuzzy Econ Rev 24(2)
14. Espinoza-Audelo LF, Leon-Castro E, Mellado-Cid C, Merigo JM, Blanco-Mesa F (2021) Uncertain induced prioritized aggregation operators in the analysis of the imports and exports. J Mult-Valued Log Soft Comput 36(6)
15. Fullér R, Majlender P (2001) An analytic approach for obtaining maximal entropy OWA operator weights. Fuzzy Sets Syst 124(1):53–57
16. Glasserman P, Xu X (2014) Robust risk measurement and model risk. Quant Financ 14(1):29–58

17. Jallal MA, Gonzalez-Vidal A, Skarmeta AF, Chabaa S, Zeroual A (2020) A hybrid neuro-fuzzy inference system-based algorithm for time series forecasting applied to energy consumption prediction. Appl Energy 268:114977
18. Jang J-S (1993) ANFIS: adaptive-network-based fuzzy inference system. IEEE Trans Syst, Man, Cybern 23(3):665–685
19. Jilani TA, Burney SMA (2008) A refined fuzzy time series model for stock market forecasting. Physica A 387(12):2857–2862
20. Kacprzyk J, Yager RR, Merigo JM (2019) Towards human-centric aggregation via ordered weighted aggregation operators and linguistic data summaries: a new perspective on zadeh's inspirations. IEEE Comput Intell Mag 14(1):16–30
21. Kim T, Shin J-Y, Kim H, Kim S, Heo J-H (2019) The use of large-scale climate indices in monthly reservoir inflow forecasting and its application on time series and artificial intelligence models. Water 11(2):374
22. Leon-Castro E, Espinoza-Audelo LF, Aviles-Ochoa E, Merigo JM, Kacprzyk J (2019) A new measure of volatility using induced heavy moving averages. Technol Econ Dev Econ 25(4):576–599
23. Leung K, Choy KL, Ho GT, Lee CK, Lam H, Luk C (2019) Prediction of B2C e-commerce order arrival using hybrid autoregressive-adaptive neuro-fuzzy inference system (AR-ANFIS) for managing fluctuation of throughput in e-fulfilment centres. Expert Syst Appl 134:304–324
24. Luo D, Zeng S, Yu G (2020) Pythagorean fuzzy investment multiple attribute decision making method based on combined aggregation method. J Intell Fuzzy Syst 39(1):949–959
25. MacDonald R (1999) Exchange rate behaviour: Are fundamentals important? Econ J 109(459):F673–F691
26. Maqsood H, Mehmood I, Maqsood M, Yasir M, Afzal S, Aadil F, Selim MM, Muhammad K (2020) A local and global event sentiment based efficient stock exchange forecasting using deep learning. Int J Inf Manage 50:432–451
27. Merigo JM, Casanovas M (2010) The fuzzy generalized OWA operator and its application in strategic decision making. Cybern Syst: Int J 41(5):359–370
28. Merigó JM, Casanovas M (2011) Induced and uncertain heavy OWA operators. Comput Ind Eng 60(1):106–116
29. Merigo JM, Gil-Lafuente AM (2010) New decision-making techniques and their application in the selection of financial products. Inf Sci 180(11):2085–2094
30. Patel PJ, Patel NJ, Patel AR (2014) Factors affecting currency exchange rate, economical formulas and prediction models. Int J Appl Innov Eng Manag 3(3):53–56
31. Perez-Arellano LA, Blanco-Mesa F, Leon-Castro E, Alfaro-Garcia V (2021) Bonferroni prioritized aggregation operators applied to government transparency. Mathematics 9(1):24
32. Pertselakis M, Tsapatsoulis N, Kollias SD, Stafylopatis A (2003) An adaptive resource allocating neural fuzzy inference system
33. Plakandaras V, Papadimitriou T, Gogas P (2015) Forecasting daily and monthly exchange rates with machine learning techniques. J Forecast 34(7):560–573
34. Puspitasari I, Akbar MS, Lee MH (2012) Two-level seasonal model based on hybrid ARIMA-ANFIS for forecasting short-term electricity load in Indonesia, pp 1–5
35. Ramakrishnan S, Butt S, Chohan MA, Ahmad H (2017) Forecasting Malaysian exchange rate using machine learning techniques based on commodities prices, pp 1–5
36. Ren S (2022) Optimization of enterprise financial management and decision-making systems based on big data. J Math
37. Rhee BK, Kim TW (2022) Fundamentals and machine learning in forecast of won-dollar exchange rate. J Korean Data Anal Soc (JKDAS) 24(6):2099–2113
38. Sanders NR, Ritzman LP (1992) The need for contextual and technical knowledge in judgmental forecasting. J Behav Decis Mak 5(1):39–52
39. Sarıca B, Eğrioğlu E, Aşıkgil B (2018) A new hybrid method for time series forecasting: AR–ANFIS. Neural Comput Appl 29:749–760
40. Sun S, Wang S, Wei Y (2020) A new ensemble deep learning approach for exchange rates forecasting and trading. Adv Eng Inform 46:101160

41. Tanaka-Yamawaki M, Tokuoka S (2007) Adaptive use of technical indicators for the prediction of intra-day stock prices. Phys A: Stat Mech Its Appl 383(1):125–133
42. Yager RR (1988) On ordered weighted averaging aggregation operators in multicriteria decisionmaking. IEEE Trans Syst Man Cybern 18(1):183–190
43. Yager RR (2002) Heavy OWA operators. Fuzzy Optim Decis Making 1(4):379–397
44. Yager RR (2008) Prioritized aggregation operators. Int J Approx Reason 48(1):263–274
45. Zhang GP (2003) Time series forecasting using a hybrid ARIMA and neural network model. Neurocomputing 50:159–175
46. Zhang Y, Hamori S (2020) The predictability of the exchange rate when combining machine learning and fundamental models. J Risk Financ Manag 13(3):48

Chapter 13
Fuzzy Risk Analysis Assessment Applied to Value Stream Mapping

Luis Pérez-Domínguez, Pedro Ángel García Aguirre, Delia-Julieta Valles-Rosales, David Luviano-Cruz, and Erwin Martínez Gómez

1 Introduction

The Outsourcing Process (OP) is nowadays used by the industrial corporations with the intention to simplify their internal manufacturing activities, and to get the expected revenue while a third party manufactures their products [1]. Risk for customers and new products are also important aspects to consider during the OP process [2], likewise, the recent global COVID-19 situation is driving the firms to use OP [3]; in addition, global companies look for OP as a business model [4]. Though OP represents several risks because the intervention of different processes and cross-functional areas are interconnected [6], adding new risks for the firms to manage, making them to look for new Risk Analysis (RA) tools to mitigate the risks and archive the projected goals [7, 8].

Failure Mode and Effect Analysis (FMEA) is a potent to handle the risks in order to perform an extended RA. The FMEA method is also a frequently employed technique throughout the OP [6]. Nevertheless, FMEA attempts to define to the RA since the human ultimate judgment on what the risks are [9–11]. Likewise, diverse studies are screening how to handle the vagueness of the FMEA with Multicriteria methods [12–16]. Moreover, Pythagorean MCDM-Fuzzy risk analysis, [17], is a potential and true strategy for addressing ambiguity during RA. Furthermore, it adds the advantage to allow the stakeholders take better decisions before the OP execution, meanwhile improving where to allocate the budget using just the required resources at the right process step, and risk is identified.

L. Pérez-Domínguez (✉) · P. Á. García Aguirre · D. Luviano-Cruz · E. M. Gómez
Departamento de ingeniería Industrial y Manufactura, Universidad Autónoma de Ciudad Juárez, Ciudad Juárez, Chihuahua, Mexico
e-mail: luis.dominguez@uacj.mx

D.-J. Valles-Rosales
Department Chair-Industrial Management & Technology, Frank H. Dotterweich-College of Engineering Texas A&M University-Kingsville, A.F. Gross Hall, MSC 203,700 University Dr. Kingsville, 78363 Kingsville, Texas, USA

M. Sahni et al. (eds.), *Novel Developments in Futuristic AI-based Technologies*, Algorithms for Intelligent Systems, https://doi.org/10.1007/978-981-99-3076-0_13

This document presents a real-life case taken from an industrial company. In this mode, the PFDA-FMEA-VSM method is used to perform a RA before manufacturing OP. Furthermore, a comparison between conventional FMEA and PFDA-FMEA-VSM analyzes the differences.

The rest of this document is organized as follows. Section 2 describes the basic concepts required to apply the PFDA-FMEA-VSM method. Then Sect. 3 contains the real-case scenario, applying the PFDA-FMEA-VSM method in a manufacturing company for an OP. Later, Sect. 4 covers the results and discussion. Finally, Sect. 5 includes the conclusion of this work.

2 Basic Concepts

This section discusses the fundamental principles involved in delivering the PFDA-FMEA-VSM approach.

Definition 1 The SMEt should be created by the project leader considering at least three experts, desirable from different functional areas.

Definition 2 SME weights (SMEw) assignation refers to a value determined for each SME associate, based on their OP know-how. The sum of the SME weights must be equal to 1. It is suggested that the project leader assign the SME weights.

During this exercise, the SMEt is formed by three senior managers from different functional areas; since the three experts have similar experience in OP, the SMEw are divided equally.

The PFDA-FMEA-VSM method is originally suggested for new product development process risk management, then to allow the adjustment of this methodology to a manufacturing OP, it is required to use next analogies.

- Phase 1 represents the period between planning stage to the kickoff meeting
- Phase 2 is the initial segment of the project
- Phase 3 symbolizes the project implementation.

3 PFDA-FMEA-VSM Application

This section shows the steps to complete the PFDA-FMEA-VSM method application for an OP in a manufacturing company. Figure 1 depicts the main steps followed to apply the PFDA-FMEA-VSM methodology.

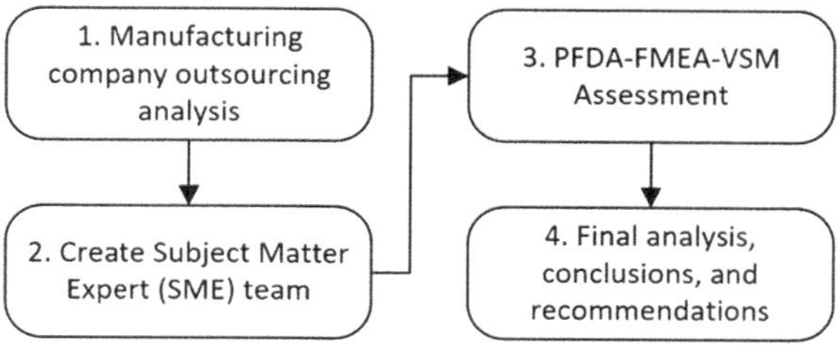

Fig. 1 PFDA-FMEA-VSM application summary diagram

4 Results and Discussion

This section shows the PFDA-FMEA-VSM application result, along with the comparison between the conventional FMEA analysis and the PFDA-FMEA-VSM analysis; this comparison is just for reference and to remark the advantages of the PFDA-FMEA-VSM method. Table 1 shows the list of the risks identified during the SMEt assessment.

Likewise, Fig. 2 illustrates the VSM current state (Table 2).

Following the VSM, future state is executed. SMEt agreed to select the top 15 risks as potential threads, highlighted in Fig. 3.

Moreover, it allows stakeholders to comprehend the main risks in a thorough diagram, making it easier to make better decisions on where to allocate resources and reduce risks in OP. This document depicts a real-world case situation in a manufacturing company using the PFDA-FMEA-VSM technique.

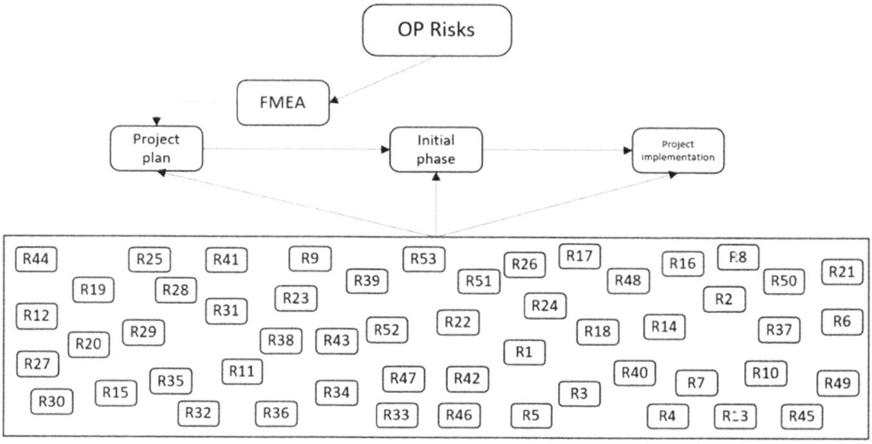

Fig. 2 OP VSM current scenario

Table 1 Potential failure modes (risks) identified by the SMEt

ID	Risk	ID	Risk	ID	Risk
1	Schedule Execution issues	19	Raw material long leadtime	36	Confidential information in risk
2	Continuous improvement versus targets	20	Lack of administrative resources	37	Raw material obsolescence not identified
3	People turnover	21	Slow hiring process	38	Single manufacturing source
4	Product validation	22	People not hired on time	39	Lack of manufacturing space because of budget
5	Supplier capacity issues	23	Lack of right equipment	40	Process capability issues
6	Sales projection inaccurate	24	People turnover during transition	41	Schedule execution issues
7	Product changes not communicated	25	Poor process documentation	42	Poor infrastructure at supplier
8	Poor supplier quality	26	Production forecast not well communicated	43	Poor engineering changes implementation
9	Manufacturing issues not communicated	27	Product quality does not meet prior transition	44	People not hired on time
10	Manufacturing errors, bad execution	28	Poor raw material management	45	Supplier decommit
11	Poor engineering changes implementation	29	External agency approval long leadtime	46	Supplier lack of capacity
12	Project scope changes	30	Poor communication customer-supplier	47	Logistics issues
13	KPI bad results from supplier	31	Missing information during transition	48	Manufacturing certification issues
14	Lack of right resources assigned	32	Poor training on new processes	49	ERP system issues
15	Lack of product history tracking at supplier	33	Poor transition product information	50	Lack of expertise on manufacturing services
16	Product specs not shared	34	Poor knowledge transfer	51	Financial issues (supplier)
17	Slow response to peak of demand	35	Administrative resources not properly assigned	52	Lack of administrative resources
18	Raw material long leadtime			53	Project transition delay

For comparison proposes, a conventional FMEA was performed by the same SMEt; Table 3 depicts the FMEA results. SMEt is considered to mitigate any risk above RPN of 25.

Conventional FMEA assessment shows 31 identified risks above 25 RPN value, while PFDA-FMEA-VSM shows top 15 rankings capturing the main risks to consider as potential real threats for the OP project. Using conventional FMEA method, all risks with RPN above 25 should have a mitigation recommended activity, besides, PFDA-FMEA-VSM top 15 optimize the resources in the mitigation process, just by

Table 2 FMEA versus PFDA-FMEA-VSM ranking comparison

ID	Risk	S	O	D	RPN	PFDA-FMEA-VSM Ranking	ID	Risk	S	O	D	RPN	PFDA-FMEA-VSM Ranking	ID	Risk	S	O	D	RPN	PFDA-FMEA-VSM Ranking
1	Schedule Execution issues	5	5	3	75	17	19	Raw material long leadtime	5	5	1	25	49	36	Confidential information in risk	5	3	1	15	1
2	Continuous improvement versus targets	4	2	4	32	13	20	Lack of administrative resources	5	5	1	25	3	37	Raw material obsolescence not identified	1	3	3	9	33
3	People turnover	5	3	5	75	37	21	Slow hiring process	5	5	1	25	3	38	Single manu-facturing source	3	3	1	9	41
4	Product validation	5	5	3	75	35	22	People not hired on time	1	2	3	6	16	39	Lack of man-ufacturing space because of budget	1	3	3	9	18
5	Supplier capacity issues	5	5	3	75	25	23	Lack of right equipment	1	2	3	6	6	40	Process capability issues	4	2	4	32	45
6	Sales projection inaccurate	5	5	3	75	40	24	People turnover during transition	1	2	3	6	31	41	Schedule execution issues	4	2	4	32	42
7	Product changes not communi-cated	5	3	3	45	42	25	Poor process documenta-tion	5	3	1	15	7	42	Poor infrastructure at supplier	5	1	1	5	22
8	Poor supplier quality	5	3	3	45	48	26	Production forecast not well commu-nicated	5	2	3	30	27	43	Poor engineering changes implementa-tion	3	1	1	3	34

(continued)

Table 2 (continued)

ID	Risk	S	O	D	RPN	PFDA-FMEA-VSM Ranking	ID	Risk	S	O	D	RPN	PFDA-FMEA-VSM Ranking	ID	Risk	S	O	D	RPN	PFDA-FMEA-VSM Ranking
9	Manufacturing issues not communicated	3	2	3	18	25	27	Product quality not meet prior transition	1	5	3	15	8	44	People not hired on time	1	3	1	3	32
10	Manufacturing errors, bad execution	3	2	3	18	8	28	Poor raw material management	3	1	5	15	36	45	Supplier decommit	5	3	2	30	51
11	Poor engineering changes implementation	3	3	3	27	18	29	External agency approval long leadtime	1	1	1	1	12	46	Supplier lack of capacity	5	3	2	30	21
12	Project scope changes	3	3	3	27	46	30	Poor communication customer-supplier	5	1	3	15	1	47	Logistics issues	5	3	2	30	28
13	KPI bad results from supplier	5	5	1	25	47	31	Missing information during transition	5	3	1	15	11	48	Manufacturing certification issues	5	3	2	30	20
14	Lack of right resources assigned	5	5	2	50	5	32	Poor training on new processes	5	3	1	15	38	49	ERP system issues	5	3	2	30	39
15	Lack of product history tracking at supplier	5	5	1	25	10	33	Poor transition product information	3	5	1	15	30	50	Lack of expertise on manufacturing services	5	3	2	30	14

(continued)

Table 2 (continued)

ID	Risk	S	O	D	RPN	PFDA-FMEA-VSM Ranking	ID	Risk	S	O	D	RPN	PFDA-FMEA-VSM Ranking	ID	Risk	S	O	D	RPN	PFDA-FMEA-VSM Ranking
16	Product specs not shared	5	5	1	25	24	34	Poor knowledge transfer	5	3	1	15	15	51	Financial issues (supplier)	5	3	2	30	29
17	Slow response to peak of demand	5	5	1	25	52	35	Administrative resources not properly assigned	5	3	1	15	23	52	Lack of administrative resources	5	3	2	30	44
18	Raw material long leadtime	5	5	1	25	50								53	Project transition delay	5	3	2	30	53

Table 3 Conventional FMEA assessment by SME

ID	Risk	S	O	D	RPN	ID	Risk	S	O	D	RPN	Risk	S	O	D	RPN	Risk
1	Schedule Execution issues	5	5	3	75	19	Raw material long leadtime	5	5	1	25	36	Confidential information in risk	5	3	1	15
2	Continuous improvement versus targets	4	2	4	32	20	Lack of administrative resources	5	5	1	25	37	Raw material obsolescence not identified	1	3	3	9
3	People turnover	5	3	5	75	21	Slow hiring process	5	5	1	25	38	Single manufacturing source	3	3	1	9
4	Product validation	5	5	3	75	22	People not hired on time	1	2	3	6	39	Lack of manufacturing space because of budget	1	3	3	9
5	Supplier capacity issues	5	5	3	75	23	Lack of right equipment	1	2	3	6	40	Process capability issues	4	2	4	32
6	Sales projection inaccurate	5	5	3	75	24	People turnover during transition	1	2	3	6	41	Schedule execution issues	4	2	4	32
7	Product changes not communicated	5	3	3	45	25	Poor process documentation	5	3	1	15	42	Poor infrastructure at supplier	5	1	1	5

(continued)

Table 3 (continued)

ID	Risk	S	O	D	RPN	ID	Risk	S	O	D	RPN	Risk	S	O	D	RPN	Risk
8	Poor supplier quality	5	3	3	45	26	Production forecast not well commu-nicated	5	2	3	30	43	Poor engineering changes implementa-tion	3	1	1	3
9	Manufacturing issues not communicated	3	2	3	18	27	Product quality does not meet prior transition	1	5	3	15	44	People not hired on time	1	3	1	3
10	Manufacturing errors, bad execution	3	2	3	18	28	Poor raw material management	3	1	5	15	45	Supplier decommit	5	3	2	30
11	Poor engineering changes implementation	3	3	3	27	29	External agency approval long leadtime	1	1	1	1	46	Supplier lack of capacity	5	3	2	30
12	Project scope changes	3	3	3	27	30	Poor commu-nication customer-supplier	5	1	3	15	47	Logistics issues	5	3	2	30
13	KPI bad results from supplier	5	5	1	25	31	Missing information during transition	5	3	1	15	48	Manufacturing certification issues	5	3	2	30
14	Lack of right resources assigned	5	5	2	50	32	Poor training on new processes	5	3	1	15	49	ERP system issues	5	3	2	30
15	Lack of product history tracking at supplier	5	5	1	25	33	Poor transition product information	3	5	1	15	50	Lack of expertise on manufacturing services	5	3	2	30

(continued)

Table 3 (continued)

ID	Risk	S	O	D	RPN	ID	Risk	S	O	D	RPN	Risk	S	O	D	RPN	Risk
16	Product specs not shared	5	5	1	25	34	Poor knowledge transfer	5	3	1	15	51	Financial issues (supplier)	5	3	2	30
17	Slow response to peak of demand	5	5	1	25	35	Administrative resources not properly assigned	5	3	1	15	52	Lack of administrative resources	5	3	2	30
18	Raw material long leadtime	5	5	1	25							53	Project transition delay	5	3	2	30

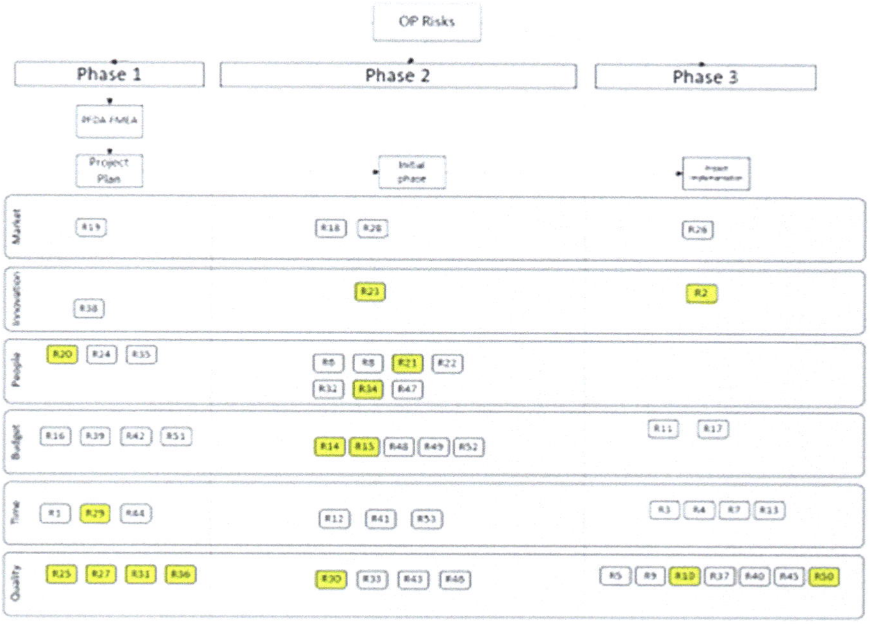

Fig. 3 PFDA-FMEA-VSM future state chart result

applying preventive methods where and when required. Table 4 reveals the comparison between PFDA-FMEA-VSM and FMEA.

5 Conclusions

A recurrent issue identifying risks in OP is the uncertainty added by the human intervention ranking the risks; moreover, by using the PFDA-FMEA-VSM method, this problem is solved and improves OP with significant advantages over the conventional FMEA. Following is a list of the primary benefits of using PFDA-FMEA-VSM for OP.

- Ranking uncertainty manipulated with Pythagorean Fuzzy Sets
- Clear visibility on the risks to be mitigated
- Optimize resources by mitigating just the major risks
- Visual risk identification, where and when is the risk.

The PFDA-FMEA-VSM method was at first used to a new product introduction process; moreover, this application reveals that it is well adapted to OP making clear and easier the OP. Likewise, there is a value added using this method, because of the risk classification by area and the project period.

Table 4 FMEA versus PFDA-FMEA-VSM ranking comparison

ID	Risk	S	O	D	RPN	PFDA-FMEA-VSM Ranking	ID	Risk	S	O		RPN	PFDA-FMEA-VSM Ranking	ID	Risk	S	O	D	RPN	PFDA-FMEA-VSM Ranking
1	Schedule Execution issues	5	5	3	75	17	19	Raw material long leadtime	5	5	1	25	49	36	Confidential information in risk	5	3	1	15	1
2	Continuous improvement versus targets	4	2	4	32	13	20	Lack of administrative resources	5	5	1	25	3	37	Raw material obsolescence not identified	1	3	3	9	33
3	People turnover	5	3	5	75	37	21	Slow hiring process	5	5	1	25	3	38	Single manufacturing source	3	3	1	9	41
4	Product validation	5	5	3	75	35	22	People not hired on time	1	2	3	6	16	39	Lack of manufacturing space because of budget	1	3	3	9	18
5	Supplier capacity issues	5	5	3	75	25	23	Lack of right equipment	1	2	3	6	6	40	Process capability issues	4	2	4	32	45
6	Sales projection inaccurate	5	5	3	75	40	24	People turnover during transition	1	2	3	6	31	41	Schedule execution issues	4	2	4	32	42
7	Product changes not communi-cated	5	3	3	45	42	25	Poor process documenta-tion	5	3	1	15	7	42	Poor infrastructure at supplier	5	1	1	5	22

(continued)

Table 4 (continued)

ID	Risk	S	O	D	RPN	PFDA-FMEA-VSM Ranking	ID	Risk	S	O		RPN	PFDA-FMEA-VSM Ranking	ID	Risk	S	O	D	RPN	PFDA-FMEA-VSM Ranking
8	Poor supplier quality	5	3	3	45	48	26	Production forecast not well communicated	5	2	3	30	27	43	Poor engineering changes implementation	3	1	1	3	34
9	Manufacturing issues not communicated	3	2	3	18	25	27	Product quality not meet prior transition	1	5	3	15	8	44	People not hired on time	1	3	1	3	32
10	Manufacturing errors, bad execution	3	2	3	18	8	28	Poor raw material management	3	1	5	15	36	45	Supplier decommit	5	3	2	30	51
11	Poor engineering changes implementation	3	3	3	27	18	29	External agency approval long leadtime	1	1	1	1	12	46	Supplier lack of capacity	5	3	2	30	21
12	Project scope changes	3	3	3	27	46	30	Poor communication customer-supplier	5	1	3	15	1	47	Logistics issues	5	3	2	30	28
13	KPI bad results from supplier	5	5	1	25	47	31	Missing information during transition	5	3	1	15	11	48	Manufacturing certification issues	5	3	2	30	20
14	Lack of right resources assigned	5	5	2	50	5	32	Poor training on new processes	5	3	1	15	38	49	ERP system issues	5	3	2	30	39

(continued)

Table 4 (continued)

ID	Risk	S	O	D	RPN	PFDA-FMEA-VSM Ranking	ID	Risk	S	O		RPN	PFDA-FMEA-VSM Ranking	ID	Risk	S	O	D	RPN	PFDA-FMEA-VSM Ranking
15	Lack of product history tracking at supplier	5	5	1	25	10	33	Poor transition product information	3	5	1	15	30	50	Lack of expertise on manufacturing services	5	3	2	30	14
16	Product specs not shared	5	5	1	25	24	34	Poor knowledge transfer	5	3	1	15	15	51	Financial issues (supplier)	5	3	2	30	29
17	Slow response to peak of demand	5	5	1	25	52	35	Administrative resources not properly assigned	5	3	1	15	23	52	Lack of administrative resources	5	3	2	30	44
18	Raw material long leadtime	5	5	1	25	50								53	Project transition delay	5	3	2	30	53

Furthermore, future works are considered by applying and adapting PFDA-FMEA-VSM to other processes, as well as trying to automate the process using a programmed software.

Acknowledgements This research was funded by CONACYT and IIT-UACJ.

References

1. Huang H, Liu F, Zhang P (2021) To outsource or not to outsource? Warranty service provision strategies considering competition, costs and reliability. Int J Prod Econ 242:108298
2. Good V, Calantone R (2019) When to outsource the sales force for new products. Ind Mark Manag 82:106–116
3. Tsai C, Ho T, Lin J, Tu C, Chang C (2021) Model for evaluating outsourcing logistics companies in the COVID-19 pandemic. Logistics 5:64
4. Tsai C, Ho T, Lin J, Tu C, Chang C (2021) Model for evaluating outsourcing logistics companies in the COVID-19 pandemic. Logistics 5:64
5. Strese S, Meuer M, Flatten T, Brettel M (2016) Organizational antecedents of cross-functional coopetition: The impact of leadership and organizational structure on cross-functional coopetition. Ind Mark Manag 53:42–55
6. Liu P, Wang S (2020) Evolutionary game analysis of cold chain logistics outsourcing of fresh food enterprises with operating risks. IEEE Access 8:127094–127103
7. Liu J, Wang D, Lin Q, Deng M (2023) Risk assessment based on FMEA combining DEA and cloud model: a case application in robot-assisted rehabilitation. Expert Syst Appl 214:119119
8. Liu Z, Wei Z, Fang Y (2023) FMEA assessment under heterogeneous hesitant fuzzy preference relations: based on extended multiplicative consistency and group decision making. IEEE Access 11:5246–5266
9. Liu Z, Xiao F (2019) An intuitionistic evidential method for weight determination in FMEA based on belief entropy. Entropy 21:211
10. Liu J, Wang D, Lin Q, Deng M (2023) Risk assessment based on FMEA combining DEA and cloud model: A case application in robot-assisted rehabilitation. Expert Syst Appl 214:119119
11. Oehmen J, Guenther A, Herrmann J, Schulte J, Willumsen P (2020) Risk management in product development: risk identification, assessment, and mitigation-a literature review. Proc Des Soc: Des Conf 1:657–666
12. Du J, Liu S, Tao L, Dong W (2023) Three-way failure mode and effect analysis approach for reliability management in multigranular unbalanced linguistic contexts. Comput Ind Eng 175:108909
13. Wang Y, Chin K, Poon G, Yang J (2009) Risk evaluation in failure mode and effects analysis using fuzzy weighted geometric mean. Expert Syst Appl 36:1195–1207
14. Hua Z, Jing X, Martínez L (2023) An ELICIT information-based ORESTE method for failure mode and effect analysis considering risk correlation with GRA-DEMATEL. Inf Fusion 393:396–411
15. Yu J, Zeng Q, Yu Y, Wu S, Ding H, Ma W, Gao H, Yang J (2023) Failure mode and effects analysis based on rough cloud model and MULTIMOORA method: Application to single-point mooring system. Appl Soft Comput 132:109841
16. Akram M, Luqman A, Alcantud J (2021) Risk evaluation in failure modes and effects analysis: hybrid TOPSIS and ELECTRE I solutions with Pythagorean fuzzy information. Neural Comput Appl 33:5675–5703

17. Aguirre P, Pérez-Domínguez L, Luviano-Cruz D, Gó mez E, Olguin I, Ramírez J (2021) Risk assessment with value added Pythagorean fuzzy failure mode and effect analysis for stakeholders. IEEE Access 9:149560–149568
18. Kumari S, Ahmad K, Khan Z, Ahmad S (2023) Failure mode and effects analysis of common effluent treatment plants of humid sub-tropical regions using fuzzy based MCDM methods. Eng Fail Anal 145:107010
19. Aguirre P, Pérez-Domínguez L, Luviano-Cruz D, Gómez E, Olguin I, Ramírez J (2021) Risk assessment with value added Pythagorean fuzzy failure mode and effect analysis for stakeholders. IEEE Access 9:149560–149568
20. Zaher M, Eldakhly N (2023) Integrated Multi-Criteria Decision Making Via Trapezoidal Neutrosophic Sets to Evaluate the Risks of Failure Mode. J Neutrosophic Fuzzy Syst (JNFS) 5:23–29

Chapter 14
Classification of the Population Health Status with the MR-Sort Method

Veronica Larreta, Pavel Anselmo Alvarez, Tania Gadea, Alejandra Duarte, and Ernesto León-Castro

1 Introduction

Health is a public good of vital importance; it is a determinant of the well-being of society. The population's health plays a fundamental role in countries' economic growth, social welfare and human development [12]. The World Health Organization indicates that it is a priority for governments that the population's health is the centre of their development and growth plans. Yet, globally, only three percent of health budgets are spent on promotion and prevention to generate well-being and reduce population morbidity and mortality [35].

V. Larreta · P. A. Alvarez (✉)
Department of Management and Economic Sciences, Universidad Autónoma de Occidente, 80020 Culiacan, Mexico
e-mail: pavel.alvarez@uadeo.mx

V. Larreta
e-mail: veronica.larreta@uadeo.mx

P. A. Alvarez
Decision-Making in Social Problems Research Group, Universidad Autónoma de Occidente, Culiacan, Mexico

T. Gadea
School of Medicine, Universidad Autónoma de Sinaloa, 80030 Culiacan, Mexico
e-mail: dragadea@hotmail.com

A. Duarte
Department of Engineering and Technology, Universidad Autónoma de Occidente, 80020 Culiacan, Mexico
e-mail: alejandra.duarte@uadeo.mx

E. León-Castro
Unidad Navojoa, Instituto Tecnológico de Sonora, 85860 Navojoa, Mexico
e-mail: ernesto.leon243151@potros.itson.edu.mx

M. Sahni et al. (eds.), *Novel Developments in Futuristic AI-based Technologies*, Algorithms for Intelligent Systems, https://doi.org/10.1007/978-981-99-3076-0_14

Morbidity refers to reported infectious diseases or chronic, genetic and other conditions. It is valid statistical data to study the processes of evolution and control of diseases [33]. Likewise, [34] defines mortality as the basic cause of death, the illness or injury that triggered the sequence of pathological events that directly led to death, or the circumstances of the accident or act of violence that produced the fatal injury.

The population's health status is affected by many social and economic factors, so it is generally described by epidemiological indicators such as mortality and morbidity rates [33]. These data are essential for decision-making during the planning process of health programmes. In addition, they have the key function of early warning of possible global, national and regional public health threats [22, 30, 36].

However, in this regard, there are no basic statistics in many countries, generally due to the lack of resources to establish systematic registries of vital statistics [5]. Therefore, the lack of timely information on the morbidity and mortality of the population limits the implementation of preventive and therapeutic measures that prevent the development of the disease or complications that cause the patient's death [22]. For this reason, the availability of accurate data on the population's health status is an obvious priority for health services.

In Mexico, the health secretary indicated an increase in the prevalence and mortality of diseases in the population, affecting more seriously the population in a condition of vulnerability [29]. At present, it is a priority for governments and departments of health to improve health protection under a comprehensive approach that prioritizes prevention and awareness of health risks and the timely treatment and control of diseases, especially those that represent the most significant impact on mortality of the population [9].

In the Sinaloa region of Mexico, 15 projects with a high impact on health are being developed. They aim to reduce morbidity and mortality rates through a situational diagnosis where there is a risk to health [13]. The analysis of health problems represents a more significant economic and social burden for the state due to the lifestyle of the population, the lack of a healthy diet, the high consumption of industrialized foods, and a sedentary lifestyle affect the increase in morbidities suffered at an early age and that lead to a low quality of life [24].

This paper analyses the morbidity and mortality of 18 municipalities in Sinaloa, Mexico. It is addressed as multicriteria sorting problem with a sorting method (ordered classification) called MR-sort. The data were obtained from the National System of Basic Information on Health [30]. The main contribution of the research is identifying the population's health status based on the prevalence of morbidity and mortality in the state.

2 Literature Review

This section reviews the literature on the Multi-Criteria Decision-Making Analysis methodologies applied to study various health problems.

In [18], it is applied TOPSIS to solve the prioritization of surgical treatments for early-stage gastric cancer. In addition, [23] have implemented ELECTRE III method to help a patient choose the optimal therapeutic programme and therapeutic schedules for brain-metastasized non-small cell lung cancer (NSCLC). Additionally, [8] used AHP, ELECTRE II, TOPSIS and PROMETHEE II methods to classify the departments and improve the appropriate choice of the patient's room in the process of hospitalization should be assigned.

Likewise, [14] proposed a mod composed of the AHP and TOPSIS methods to choose the optimal date of return to school for university students due to COVID-19 disease risks. On the other hand, [28], through the AHP method, studied the level of willingness of the population to apply the COVID-19 vaccine in Pakistan.

Furthermore, [19] PROMETHEE II method developed a model for prioritizing municipalities to monitor chronic Chagas disease. Likewise, [32] used the PROMETHEE II method to identify how different groups prioritize care for sickle cell disease among three other diseases included in the Strategic Component of Pharmaceutical Services.

As well as [25] applied COPRAS method for assessing and identifying the potential influence of socioeconomic risk factors on non-communicable disease mortality. As well as, [25] applied the G-TOPSIS method to identify treatments, set priorities and effectively allocate resources to minimize mortality for disease cardiac. Besides, [10] combined features of the outranking ELECTRE method, the reference level TOPSIS method and fuzzy logic analysing how Mycobacterium Tuberculosis are risk factors of deaths.

On the other hand, the public health problem can be associated with different phenomena. Roldán et al. [27] analysed the public health in Mexico regarding some degrees of marginalization and health service coverage. In Mexico, diversity of studies have been performed around the marginalization as [1, 2, 21]. The first studied the Sinaloa region in Mexico for marginalization in a multicriteria ranking problem. The second analyses the marginalization of regions of Mexico with a multicriteria sorting method. The third generates a robust multicriteria ranking of marginalized regions of Mexico.

3 Methodology

The process of assign alternatives to classes that are ordered from the best to the worst corresponds to the decision-making problem called sorting. The multicriteria decision-making sorting consists of assigning a set of alternatives $A = \{a_1, a_2, \ldots, a_m\}$ in predefined ordered classes $C_1, C_2, \ldots, C_k$, where the class C_h is preferred to class C_l ($C_h \succ C_l$). Each alternative a_i is evaluated by criteria in $G = \{g_1, g_2, \ldots, g_n\}$ and compared against a limit profile or central profile. The limiting profile lp is the vector of attributes values $(b_{h,1}, \ldots, b_{h,j}, \ldots, b_{h,n})$. The central profile cp is the vector of attributes values of a centroid representing the class

$(cp_{h,1}, \ldots, cp_{h,j}, \ldots, cp_{h,k})$. In this sense, the classes are represented by limiting or central profiles.

The MR-sort is the Sorting Method based on a Majority Rule. It was proposed by [6, 7] as a noncompensatory sorting method. Moreover, it is based on the ELECTRE-TRI method. The description of the method is presented as follows based on the notation from [17, 31].

The class C_h in the MR-sort method is delimited by two limiting profiles, b_{h-1} and b_h, the lower and upper limiting profiles, respectively. An alternative a_i is compared against the lower and upper limiting profiles of category C_h. If a_i is at least as good as the category's lower profile b_{h-1} and that condition is not fulfilled when a_i is compared to the category's upper profile b_h then a_i is assigned to a category C_h. If there is majority of criteria ($\sum_{g_j(a_i) \geq b_{h-,j}} w_j$) supporting the assertion a_i is at least as good as the profile b_h, $\sigma(a_i \succ b_{h-1})$, a_i is assigned to a category C_h.

$$\sum_{\substack{j \in N: \\ g_j(a_i) \geq b_{h-1,j}}} w_j \geq \lambda \ \ and \ \sum_{\substack{j \in N: \\ g_j(a_i) \geq b_{h,j}}} w_j < \lambda \tag{1}$$

where $N = \{1, \ldots, n\}$, and it is assumed that $b_{h-1,j} \leq b_{h,j}$, $h = 1, \ldots, k; h < k$.

The λ is a majority threshold indicating if the coalition of criteria is strong enough to assert indeed a_i is at least as good as b_h, $\sigma(a_i \succ b_{h-1})$.

Outranking approach includes very well accepted methods in the scientific community and applied to a diversity of MCDM problems. Alvarez et al. [1] analysed municipal districts to construct a water supply system. López et al. [20], a group of decision-makers, evaluated the technological packages available to an agricultural company to select the most appropriate one. Alvarez et al. [4] studied the problem of the innovation capacity of Mexican regions with an outranking approach.

4 Evaluation of Population Health Status

This section describes the data, preferences information and results of the classification of municipalities in the predefined health ordered classes.

4.1 Data Description

In Mexico, SINBA (the acronym from the Spanish) is the health national information system that provides a set of components that integrate the activities derived from the process of health care and the provision of services that produces and disseminates statistical information [30].

The current study considers the death associated with a disease from a group of diseases prevailing in the population of a municipality. In this regard, the cause of death concerns morbidity, defined in the international classification of diseases. Those diseases are prevailing in the population, and the information is collected on the death registered in 2019 and the people's place of origin [15].

Table 1 lists the group of diseases considered for the registered death. The external causes such as accidents or other injuries are excluded. Only deaths related to the diseases in Table 1 are considered for the region's population health. This section describes the data, preferences information and results of the classification of municipalities in the predefined health ordered classes.

The region in observation is Sinaloa. It is a state in northwest Mexico divided into 18 municipalities (see Table 2). The population census by [15] shows that 3,026,943 people live in Sinaloa.

Table 3 shows the number of deaths by the municipality in each group of diseases [15]. For the classification process, the percentage of population death associated with disease was computed regarding the total population of each municipality. The aim is to compare municipalities with different population sizes. The information is

Table 1 Group of diseases as decision criteria for evaluation of population health

Label	Description
dc1	Certain infectious and parasitic diseases
dc2	Tumours
dc3	Diseases of the blood and hematopoietic organs and certain disorders that affect the mechanism of immunity
dc4	Endocrine, nutritional and metabolic diseases
dc5	Mental and behavioural disorders
dc6	Diseases of the nervous system
dc7	Diseases of the circulatory system
dc8	Diseases of the respiratory system
dc9	Diseases of the digestive system
dc10	Diseases of the skin and subcutaneous tissue
dc11	Diseases of the musculoskeletal system and tissue conjunctive
dc12	Diseases of the genitourinary system
dc13	Pregnancy, childbirth and the puerperium
dc14	Certain conditions originating in the perinatal period
dc15	Congenital malformations, deformities and anomalies chromosomal
dc16	Symptoms, signs, and abnormal clinical and diagnostic findings laboratory, not elsewhere classified

Table 2 Municipalities of Sinaloa region, Mexico

Label	Municipality
M1	Ahome
M2	Angostura
M3	Badiraguato
M4	Concordia
M5	Cosalá
M6	Culiacán
M7	Choix
M8	Elota
M9	Escuinapa
M10	El Fuerte
M11	Guasave
M12	Mazatlán
M13	Mocorito
M14	Rosario
M15	Salvador Alvarado
M16	San Ignacio
M17	Sinaloa de Leyva
M18	Navolato

from 2019 regarding the place of origin of the people. In this sense, the data is before the covid pandemic, considering a normal health world situation.

4.2 Expert Preference Information

In the current study, a healthcare expert analysed the population's health level of the Sinaloa municipalities. In addition, the healthcare expert defined the categories and parameters of the applied multicriteria decision-making sorting method.

The evaluation of the population health situation is carried on regarding three categories of health. The categories are the following: the best category is good health (C1), an intermediate category is a medium health (C2) and the worst category is bad health (C3). Those regard the population's health based on death associated with a specific disease.

The applied MR-sort method requires some parameters for the classification process. The healthcare expert participated in an elicitation process to help the analyst to define the parameters. Table 4 shows the direction of criteria, weights and limiting profiles. The criterion is defined as cost when it requires to be *minimized* and benefit criteria when it requires to be *maximized*. The weight definition is supported by the SRF method [11] based on the deck of card method, where the expert is asked to

Table 3 Deaths associated with a group of diseases

Municipality	dc1	dc2	dc3	dc4	dc5	dc6	dc7	dc8	dc9	dc10	dc11	dc12	dc13	dc14	dc15	dc16
M1	60	400	10	270	8	47	755	187	150	12	15	96	0	36	95	22
M2	9	46	1	41	0	12	75	43	11	1	1	7	0	1	4	1
M3	3	21	1	9	0	5	51	21	6	1	0	0	0	0	1	2
M4	7	26	0	20	1	2	57	14	4	1	3	3	0	1	4	4
M5	2	12	0	6	0	1	24	10	3	0	1	1	0	1	2	1
M6	163	688	22	444	19	96	1294	456	244	24	29	129	5	81	213	58
M7	5	24	0	18	0	6	65	13	10	0	1	7	0	3	7	2
M8	7	35	1	44	1	3	56	24	11	2	1	5	0	3	12	0
M9	7	51	2	44	3	8	101	29	26	0	0	14	1	7	19	3
M10	13	81	3	67	2	10	193	57	43	4	2	33	1	10	21	9
M11	44	252	6	182	12	46	479	177	101	9	10	50	1	26	71	17
M12	103	454	13	470	12	75	712	218	191	10	19	107	4	56	145	14
M13	12	50	6	39	2	9	84	43	18	1	4	4	0	0	0	2
M14	13	55	1	43	2	7	112	33	19	2	1	14	0	7	14	3
M15	7	85	1	40	4	14	107	45	24	0	6	9	0	5	16	4
M16	1	14	0	27	1	7	48	13	5	1	0	3	0	1	4	0
M17	15	70	4	53	2	14	163	52	27	3	6	9	1	2	7	6
M18	26	107	4	121	4	26	201	96	58	6	13	19	1	7	19	5

Table 4 Weights of criteria and limiting profiles

			Profiles classes	
Criteria	Min/Max	Weight	b1	b2
dc1	Min	0.0917	0.00013	0.00019
dc2	Min	0.0722	0.00087	0.00089
dc3	Min	0.0787	0.00002	0.00005
dc4	Min	0.1111	0.00059	0.00080
dc5	Min	0.0399	0.00002	0.00003
dc6	Min	0.0462	0.00010	0.00018
dc7	Min	0.1046	0.00164	0.00230
dc8	Min	0.0982	0.00041	0.00066
dc9	Min	0.0851	0.00033	0.00034
dc10	Min	0.0333	0.00003	0.00004
dc11	Min	0.0204	0.00003	0.00008
dc12	Min	0.0528	0.00012	0.00021
dc13	Min	0.0593	0.00000	0.00001
dc14	Min	0.0657	0.00003	0.00008
dc15	Min	0.0268	0.00009	0.00021
dc16	Min	0.014	0.00005	0.00008

order the criteria from the least important to the most important. The profiles were defined in an interactive process. The expert and the analyst collaborated to identify the limiting profile in each criterion between classes.

Figure 1 shows the illustration of the limiting profiles for each criterion. The values lower than the limiting profile $b_{1,j}$ indicate the municipality in criterion g_j belong to the Class 1. The values between limiting profiles $b_{1,j}$ and $b_{2,j}$ indicate the municipality in criterion g_j belong to the Class 2. The values higher than the limiting profile $b_{2,j}$ indicate the municipality in criterion g_j belong to the Class 3.

4.3 *Classification of Municipalities*

Analysing the health status of Sinaloa, 18 municipalities are assigned to a specific health category regarding the population's mortality by its morbidity. The health outcomes are shown in Table 5; here the classification of each municipality is presented. Six municipalities are assigned to the best class good health (C1), eight municipalities are assigned to intermediate class medium health (C2) and four municipalities are assigned to the worst class bad health (C3).

The municipalities in the good health category (C1) present low values of deaths in the most important criteria: Endocrine, nutritional and metabolic diseases (dc4) and

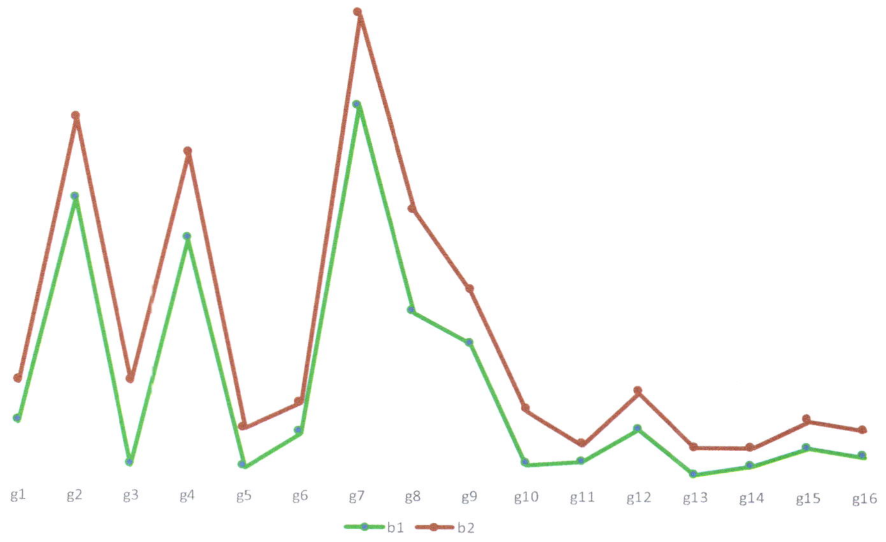

Fig. 1 Profiles of the health status categories

Table 5 Classification of municipalities based on their population health status

Label	Municipality	Category
A1	Ahome	C1
A3	Badiraguato	C1
A5	Cosalá	C1
A6	Culiacán	C1
A8	Elota	C1
A15	Salvador Alvarado	C1
A2	Angostura	C1
A4	Concordia	C1
A7	Choix	C1
A9	Escuinapa	C1
A10	El Fuerte	C1
A11	Guasave	C1
A17	Sinaloa	C1
A18	Navolato	C1
A12	Mazatlán	C1
A13	Mocorito	C1
A14	Rosario	C1
A16	San Ignacio	C1

Diseases of the circulatory system (dc7). The municipalities show population percentages between 0.00035 and 0.00080 in criterion dc4. The second most important criterion dc7 shows population percentages between 0.00101 and 0.00192.

The municipalities in the *medium health* category (C2) present higher values of deaths in dc4 and dc7. The municipalities assigned to C2 show population percentages between 0.00061 and 0.00093 in criteria dc4 and values between 0.00135 and 0.00229 in dc7.

The municipalities in the *bad health* category (C3) present the highest values of deaths in dc4 and dc7. The municipalities assigned to C3 show population percentages between 0.00082 and 0.00138 in criteria dc4, and values between 0.00142 and 0.00246 in dc7.

In this regard, the municipalities belonging to bad health category (C3) show the highest percentages of deaths in other important criteria as diseases of the respiratory system (dc8) and certain infectious and parasitic diseases (dc1). In the criterion dc8, the municipalities show the following values: Mazatlán (0.00043), Mocorito (0.00107), Rosario (0.00063) and San Ignacio (0.00067). In the same criterion, the good health category shows percentages between 0.00041 to 0.00079. In the criterion dc1, the municipalities show the following values: Mazatlán (0.00021), Mocorito (0.00030), Rosario (0.00025) and San Ignacio (0.00005). In the same criterion, the good health category shows percentages between 0.00009 and 0.00016.

The municipalities from categories C1 to C2 and C2 to C3 indicate a tendency to increase deaths in higher classes than lower classes. It means more people from a municipality belonging to class C3 are dying from a disease associated with one of the morbidities of the group of diseases from Table 1. The information regards the percentage of deaths in the total population of the municipalities.

The population's health status in Sinaloa can be related to the population's lifestyle, the lack of a healthy diet, the high consumption of industrialized foods and a sedentary lifestyle [24]. However, it requires further studies to relate to health and population activity.

5 Conclusion

The research is focused on the identification of the health status of the population of a region in Mexico. The analysis regards the prevalence of morbidity and mortality in the population. Eighteen municipalities were assigned to a population health category with the multiple criteria decision-making sorting method, MR-sort.

The outcomes obtained in the current research show the municipalities of Sinaloa with the population's highest, medium and lowest levels of health. The municipalities identified with the lowest level of health in the population present deaths associated with the primary diseases concerning the percentage of the population.

Analysing the population's health status in Sinaloa can help decision-makers of social politics develop plans and policies to reduce the deaths of the population associated with morbidities. The methodology can be applied to different regions using the data available by the government of Mexico to show the health condition of the population. In addition, the different sorting methods can be applied to compare the performance of the result reported in the current research.

Acknowledgements Research supported by Red Sistemas Inteligentes y Expertos Modelos Computacionales Iberoamericanos (SIEMCI), project number 522RT0130 in Programa Iberoamericano de Ciencia y Tecnología para el Desarrollo (CYTED) , Autonomous University of Occident in Mexico through the Institutional Program for the Strengthening of Research and Postgraduate studies (PIFIP-2022).

References

1. Alvarez PA (2022) Study of the geographical marginality in a mexican region using the MR-sort method. In: León-Castro E, Blanco-Mesa F, Alfaro-García V, Gil-Lafuente AM, Merigó JM, Kacprzyk J (eds) Soft computing and fuzzy methodologies in innovation management and sustainability. Lecture notes in networks and systems, vol 337. Springer, Cham. https://doi.org/10.1007/978-3-030-96150-3_4
2. Alvarez PA (2023) Assigning Mexican regions to categories of marginality with MR-sort. In: Sahoo L, Senapati T, Yager R (eds) Real life applications of multiple criteria decision making techniques in fuzzy domain, vol 420. Springer, Singapore. https://doi.org/10.1007/978-981-19-4929-6_28. ISBN 978–981–19–4928–9
3. Alvarez PA, Morais DC, Leyva López JC, de Almeida AT (2022) An ELECTRE III based consensus-reaching process to improve a collective solution. Int Trans Oper Res 29(2):1048–1088. https://doi.org/10.1111/itor.12798
4. Alvarez PA, Valdez C, Dutta B (2022) Analysis of the innovation capacity of Mexican regions with the multiple criteria hierarchy process. Socio-Econ Plan Sci 84. https://doi.org/10.1016/j.seps.2022.101418
5. Bonita R, Beaglehole R, Kjellström T (2008) Epidemiologia básica. OPS, Washington, D.C.
6. Bouyssou D, Marchant T (2007) An axiomatic approach to noncompensatory sorting methods in MCDM, I: the case of two categories. Eur J Oper Res 178(1):217–245. https://doi.org/10.1016/j.ejor.2006.01.027
7. Bouyssou D, Marchant T (2007) An axiomatic approach to noncompensatory sorting methods in MCDM, II: More than two categories. Eur J Oper Res 178(1):246–276. https://doi.org/10.1016/j.ejor.2006.01.033
8. Chalgham M, Khatrouch I, Masmoudi M, Walha OC, Dammak A (2019) Inpatient admission management using multiple criteria decision-making methods. Oper Res Health Care 23:100173. https://doi.org/10.1016/j.orhc.2018.10.001
9. DOF (2022) Programa sectorial derivado del plan nacional de desarrollo 2019–2024. Accessed 10 June 2022 from https://dof.gob.mx/nota_detalle.php?codigo=5598474&fecha=17/08/2020#gsc.tab=0
10. Ezhilarasan N, Felix A (2022) Fuzzy ELECTRE and TOPSIS method to analyze the risk factors of tuberculosis. J Phys: Conf Ser 2267(1):012109. https://doi.org/10.1088/1742-6596/2267/1/012109
11. Figueira J, Roy B (2002) Determining the weights of criteria in the ELECTRE type methods with a revised Simos' procedure. Eur J Oper Res 139:317–326. https://doi.org/10.1016/S0377-2217(01)00370-8
12. García RJF, García FA, Priego HO, Martínez PL (2017) Importancia de la salud para el crecimiento económico, bienestar social y desarrollo humano. ACADEMIA.
13. Gobierno del Estado de Sinaloa (2019) Tercer Informe de Gobierno. https://issuu.com/diario24horas/docs/tercer_informe_de_gobierno_2019

14. Huang G, Pan Q, Zhao S, Gao Y, Gao X (2020) Prediction of COVID-19 outbreak in china and optimal return date for university students based on propagation dynamics. J Shanghai Jiaotong Univ (Science) 25(2):140–146. https://doi.org/10.1007/s12204-020-2167-2
15. INEGI (2020) INEGI: Censo de población y vivienda 2020. Accessed 10 June 2022 from https://cuentame.inegi.org.mx/monografias/informacion/sin/poblacion/
16. INEGI_SSA (2022) Cubos dinámicos: Defunciones. Accessed 10 June 2022 from https://sinba.salud.gob.mx/cuboDEFUNCIONES/INEGI_SSA
17. Leroy A, Mousseau V, Pirlot M (2011) Learning the parameters of a multiple criteria sorting method. In: Brafman RI, Roberts FS, Tsoukias A (eds) Algorithmic decision theory, vol 6992, pp 219–233
18. Li DP, He JQ, Cheng PF, Wang JQ, Zhang HY (2018) A novel selection model of surgical treatments for early gastric cancer patients based on heterogeneous multicriteria group decision-making. Symmetry 10(6):223. https://doi.org/10.3390/sym10060223
19. Lima MM, Costa VM, da Palmeira SL, de Castro AP B (2021). Estratificação de territórios prioritários para vigilância da doença de Chagas crônica: análise multicritério para tomada de decisão em saúde. Cadernos de Saúde Pública, 37(6). https://doi.org/10.1590/0102-311x00175920
20. López, J.C.L., Carrillo, P.A.Á., Valenzuela, O.A. (2018) A multicriteria group decision model for ranking technology packages in agriculture. *Soft Computing for the Sustainability Science*. In Cruz Corona, C. (Eds.), Springer. doi: https://doi.org/10.1007/978-3-319-62359-7_7.
21. López-Parra P, Alvarez PA, Bernal M, León-Castro E (2023) A robust ranking model for analysing marginality of regions of Mexico. Int J Innovat Sustain Dev 17(1–2):104–122. https://doi.org/10.1504/IJISD.2023.127975
22. OPS (2018) Indicadores de salud. Aspectos conceptuales y operativos. https://iris.paho.org/bitstream/handle/10665.2/49058/9789275320051_spa.pdf?sequence=5&isAllowed=y
23. Pan L, Ren P, Xu Z (2018) Therapeutic Schedule Evaluation for Brain-Metastasized Non-Small Cell Lung Cancer with A Probabilistic Linguistic ELECTRE II Method. Int J Environ Res Public Health 15(9):1799. https://doi.org/10.3390/ijerph15091799
24. Plan Estatal de Desarrollo de Sinaloa (2022) Plan Estatal de Desarrollo de Sinaloa 2022–2024. https://ped.sinaloa.gob.mx/wp-content/uploads/2022/04/05-1.3-Salud-compressed.pdf
25. Rehman S, Rehman E, Hussain I, Jianglin Z (2021) Socioeconomic Influence on cardiac mortality in the South Asian region: new perspectives from grey modeling and G-TOPSIS. J Healthcare Eng 1–10. https://doi.org/10.1155/2021/6866246
26. Rehman S, Rehman N, Naz M, Mumtaz A, Jianglin Z (2021) Application of grey-based SWARA and COPRAS techniques in disease mortality risk assessment. Journal of Healthcare Engineering 2021:1–9. https://doi.org/10.1155/2021/7302157
27. Roldán J, Álvarez M, Carrasco M, Guarneros N, Ledesma J, Cuchillo-Hilario M, Chávez A (2017) Marginalization and health service coverage among Indigenous, rural, and urban populations: a public health problem in Mexico. Rural Remote Health 17(4). https://doi.org/10.22605/rrh3948
28. Sarwar A, Nazar N, Nazar N, Qadir A (2021) Measuring vaccination willingness in response to COVID-19 using a multi-criteria-decision making method. Hum Vaccin Immunother 17(12):4865–4872. https://doi.org/10.1080/21645515.2021.2004836
29. Secretaria de Salud (2019) Programa Sectorial de Salud 2019–2024. http://saludsinaloa.gob.mx/wp-content/uploads/2019/transparencia/PROGRAM_SECTORIAL_DE_SALUD_2019_2024.pdf
30. SINBA (2022) Sistemas Nacional Información Básica en Materia de Salud. Accessed 5 June 2022 from SINBA. https://sinba.salud.gob.mx/
31. Sobrie O, Mousseau V, Pirlot M (2018) Learning monotone preferences using a majority rule sorting model. Int Trans Oper Res. https://doi.org/10.1111/itor.12512
32. Souza-Filho H, Grothe R, Bueno C (2021) The sickle cell disease profile in the Brazilian public health system: a multicriteria decision analysis approach for priority setting. Hematology, Transfusion Cell Therapy 43:S30. https://doi.org/10.1016/j.htct.2021.10.052

33. Tulchinsky TH, Varavikova EA (2014) Measuring, monitoring, and evaluating the health of a population. In: The new public health. Elsevier, pp 91–147. https://doi.org/10.1016/B978-0-12-415766-8.00003-3
34. WHO. (2021). Mortality and Health. https://biblioguias.cepal.org/c.php?g=398214&p=3602022
35. WHO (2022a) Director-General's report to Member States at the 75th World Health Assembly
36. WHO (2022b) Director-General's report to Member States at the 75th World Health Assembly. https://www.who.int/director-general/speeches/detail/who-director-general-s-opening-address-at-the-75th-world-health-assembly---23-may-2022

Chapter 15
A Bibliometric Analysis of Problem Structuring Between 1980 and 2019

Kenia Sanchez-Valenzuela, Diego Alonso Gastélum Chavira, Ernesto León-Castro, and Denisse Ballardo-Cárdenas

1 Introduction

Inside organizations, decision-makers could have some new or unusual decision problems, which are unstructured. In this type of problem, information can be ambiguous or incomplete, so the decision-maker does not know the problem, which leads to the risk of solving the wrong problem [7]. Structuring a problem describes the process to reach enough understanding of the components of a problem [57]. Problem structuring (PS) is defined as a strictly specified step-by-step procedure, which can almost be said to examine a problem in terms of a previously ordered structure [55].

It is known that PS is discussed on problem-solving, creativity, and more generally in the social sciences, as well as in operations research journals [3]. It recognized the need for researching in this area as there are similar problems in other disciplines. It is because structuring the problem is the most important activity in the decision-making process to solve the correct problem [2, 4].

A Problem Structuring Method (PSM) is useful in structuring problems with multiple factors, different perspectives, partially conflicting interests, and uncertainties, which helps to focus on the precise problem to be solved, avoiding the error of the rational model [42].

K. Sanchez-Valenzuela (✉) · D. A. G. Chavira · D. Ballardo-Cárdenas
Universidad Autónoma de Occidente, Blvd. Lola Beltran S/N, 80020 Culiacán, Sin., México
e-mail: kenia.sanchez@uadeo.mx

D. A. G. Chavira
e-mail: diego.chavira@uadeo.mx

D. Ballardo-Cárdenas
e-mail: denisse.ballardo@uadeo.mx

E. León-Castro
Business School, Tecnologico de Monterrey, Campus Sinaloa, Blvd. Pedro Infante 3773 Pte, Recursos Hidraulicos, 80100 Culiacán Rosales, Sin., México
e-mail: ernesto.leon@tec.mx

M. Sahni et al. (eds.), *Novel Developments in Futuristic AI-based Technologies*, Algorithms for Intelligent Systems, https://doi.org/10.1007/978-981-99-3076-0_15

Thus, practitioners in problem structuring are interested in methods, models, methodologies, tools, and techniques to be used in the decision-making process [55]. Different methods have different purposes and, therefore, it may be necessary to cover all phases of problem structuring [16, 22]. However, from our knowledge, a bibliometric study about problem structuring has not been performed.

It is known that a bibliometric study is considered as the science that consists of carrying out the quantitative analysis of published articles [10]. It allows evaluating the impact of scientific publications using bibliometric indicators showing an overview of a specific area [35, 37]. Hence, this paper aims to provide useful information on scientific research publications in the area of Problem Structuring, from 1980 to 2019, through a bibliometric analysis. The study used a bibliometric approach to map authors, institutions, countries, and journals using the VOS visualization software [52]. Among other results, this paper shows there is a growing field of study and a considerable number of articles and citations concentrated in different magazines. The paper is organized into the following sections: Sect. 2 describes the methodology, Sect. 3 presents the results, Sect. 3.3 develops a graphical visualization of the results of using the VOS viewer software, and Sect. 4 presents the main conclusions of the paper.

2 Methodology

This analysis is performed with the bibliographic search, where the database of the main Scopus collection is used. The Scopus database has been used in different academic research as the main source [33, 50, 51]. For this study, the following process was performed:

In the first phase, search equations were built and tested from the combination of keywords used in the literature on the subject of problem structuring. As a result, the following equation was selected: problem * structuring "," problem structuring * "," decision problem structuring * "and" method of structuring problems *. For this study, this selection of keywords was labeled Problem Structuring (PS). As mentioned, it was tested on the Scopus database. In the second phase, the time period was determined in a range from 1980 to 2019.

In the third phase, it defines the types of documents. Initially, all the documents published on the subject were considered in the Scopus database, resulting in 2217 documents published during this period. Later, a more refined searching was performed, where the type of document was defined to consider only articles and reviews. Hence, the result was reduced to 1550.

Finally, In the fourth phase, a set of eight bibliometric indicators were defined. It included citations per year, citations per article, total number of citations, the total number of articles, thresholds per citation, H-index, ranking of universities in the Academic Ranking of World Universities (ARWU), and the Quacquarelli Symonds Ranking of Universities (QS). These bibliometric indicators allowed to know and understand the subject.

In order to analyze that collected data, the VOS visualization software was used. This software has been used in several studies [12, 27, 30, 43, 44], to perform bibliographic coupling, co-citation, and author keywords, through data collection and the construction of graphic maps. It is a popular tool for evaluating information on research topics [11].

3 Results

This section presents the results of this bibliometric contribution on Problem Structuring from 1980 to 2019 in Scopus database. It provides tables and figures about the topic classified by, among others, the most cited papers, leading authors, institutions, and countries.

3.1 General Perspective

According to the obtained results, the topic of Problem Structuring has been studied over the years and has shown increasing evolution, from 16 papers in 1980 to 1999 papers in 2019. This evolution is observed (see Fig. 1).

Table 1 presents the growth in the number of citations over time. It is especially observed that the study of the structuring of problems is increasing with a total of 1,550 productions and 34,371 citations. It is identified that problem structuring has been studied mainly in 2018 and 2019 according to the number of publications, but the largest number of citations is from 2006 with 2,196 and 66 publications. It is observed that since the second decade the growth is significant.

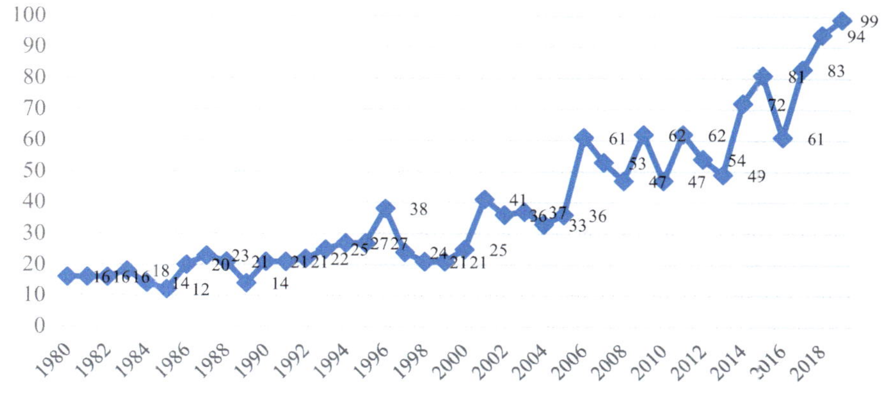

Fig. 1 The annual number of papers published in PS

Table 1 Annual citation structure of PS

Year	TP	TC	≥ 200	≥ 100	≥ 50	≥ 20	≥ 10	≥ 5	≥ 1
1980	16	261	0	0	2	3	3	2	1
1981	16	297	0	0	3	2	2	1	3
1982	16	346	0	0	3	2	0	4	2
1983	18	359	0	1	1	2	3	3	3
1984	14	161	0	0	0	3	3	3	4
1985	12	156	0	0	1	2	3	0	3
1986	20	293	0	1	0	3	6	3	4
1987	23	1354	1	1	1	4	3	4	5
1988	21	304	0	0	1	4	4	2	6
1989	14	929	1	0	2	1	3	2	4
1990	21	792	2	0	1	4	3	2	2
1991	21	239	0	0	0	5	6	3	3
1992	22	510	0	1	3	4	2	3	7
1993	25	1518	1	2	5	6	4	1	5
1994	27	701	0	0	6	5	3	6	2
1995	27	771	0	2	1	7	4	3	5
1996	38	594	0	1	3	5	1	10	12
1997	24	1015	0	5	1	5	2	3	6
1998	21	1862	4	3	1	4	2	1	2
1999	21	491	0	1	2	4	1	5	6
2000	25	1003	1	0	3	6	4	3	2
2001	41	897	0	1	4	11	8	3	6
2002	36	716	0	1	4	7	5	9	6
2003	37	665	0	1	1	12	7	2	10
2004	33	1380	2	1	6	6	5	1	3
2005	36	1235	1	1	5	6	7	7	7
2006	61	2196	2	2	8	14	11	11	8
2007	53	1407	1	2	4	14	14	3	10
2008	47	640	0	0	3	8	8	7	15
2009	62	1471	0	1	9	18	11	11	9
2010	47	1314	1	3	3	8	15	7	9
2011	62	1342	0	2	5	15	9	10	15
2012	54	1761	1	3	2	12	11	9	12
2013	49	1224	2	0	4	8	9	9	11
2014	72	1111	0	2	0	15	19	10	20
2015	81	1065	0	0	2	16	18	19	18

(continued)

Table 1 (continued)

Year	TP	TC	≥ 200	≥ 100	≥ 50	≥ 20	≥ 10	≥ 5	≥ 1
2016	61	512	0	0	1	8	8	13	20
2017	83	903	0	2	2	4	10	21	30
2018	94	414	0	0	0	3	5	24	40
2019	99	162	0	0	0	0	3	9	41
Total	1550	34,371	20	40	103	266	245	249	377
	100%		1.29%	2.58%	6.64%	14.58%	15.18%	16.06%	24.32%

Notes TP is the total papers; TC is the total of citations; the thresholds ≥200, ≥100, ≥50, ≥20, ≥10, ≥5, ≥1 are the number of papers equal or greater than 200, 100, 50, 20, 10, 5, and 1 citations, respectively

Table 2 presents the Journal's thirty-five most cited papers of all time. Note that in 1980 the first publication of problem structuring was presented. In this table, the three most cited papers are from the years 1987, 1989, and 1993.

"The Agent-structure Problem in International Relations Theory" written by Wendt A.E is in first place as the most cited article. Followed by the paper "Solution of Schrödinger's Equation for Large Systems" with a total of 719 citations written by Teter M. P., Payne M. C., et al. In third place the paper "Density-matrix Electronic-Structure Method with Linear System-size Scaling" with a total of 628 citations was written by Li X.-P., Nunes R. W., Vanderbilt D.

3.2 *Leading Authors, Institutions, and Countries*

In this section, let us analyze those authors, institutions, and countries that have the most publications in Problem Structuring. First, let us look into the 35 most productive authors see Table 3. Note that in the case of a tie, the ranking is according to the number of citations. White L. is the author that has the most publications in Problem Structuring, followed by Ackermann F. Note that both of them are among the most productive authors. Furthermore, it is worth mentioning the author Rosenhead J. of London School of United Kingdom. He is in third place of the ranking of 35 most cited.

Next, let's analyze the most productive institutions to know which universities produce the most documents and to know the universities that publish the most in the area of problem structuring. The University of Warwick is the most productive institution located in the United Kingdom with a number of 995 citations. The following institutions are the London School and the University of Strathclyde, both located in the United Kingdom as well. Likewise, it is observed in this top 35 that the country with the highest number of institutions is the United Kingdom followed by the United States.

Table 2 The most cited papers in problem structuring between 1980 and 2019

R	TC	Title	Author/s	Year	C/Y
1	937	The agent-structure problem in international relations theory	Wendt A. E	1987	29,28
2	719	Solution of Schrödingers equation for large systems	Teter M. P., Payne M. C., et al	1989	23,97
3	628	Density-matrix electronic-structure method with linear system-size scaling	Li X.-P., Nunes R. W., Vander bilt D	1993	24,15
4	558	Perspective on density functional theory	Burke K	2012	79,71
5	488	Ab initio multiple spawning: Photochemistry from first principles quantum molecular dynamics	Ben-Nun M., Quenneville J., Martínez T. J	2000	25,68
6	474	Ab initio calculations of quasiparticle band structure in correlated systems: LDA + + approach	Lichtenstein A., Katsnelson M	1998	22,57
7	472	Added-mass effect in the design of partitioned algorithms for fluid–structure problems	Causin P., Gerbeau J. F., Nobile F	2005	33,71
8	427	Multi-criteria decision analysis in natural resource management: A critical review of methods and new modelling paradigms	Mendoza G. A., Martins H	2006	32,85
9	329	Problem structuring methods in action	Mingers J., Rosenhead J	2004	21,93
10	323	Characterization of micro- and mesoporous solids by physisorption methods and pore-size analysis	Storck S., Bretinger H., Maier W. F	1998	15,38
11	322	Accurate forces in a local-orbital approach to the local-density approximation	Jackson K., Pederson M. R	1990	11,10
12	286	Analyzing cognitive maps to help structure issues or problems	Eden C	2004	19,07
13	284	A mixed displacement–pressure formulation for poroelastic materials	Atalla N., Panneton R., Debergue P	1998	13,52
14	263	Proton transfer 200 years after Von Grotthuss: Insights from ab initio simulations	Marx D	2006	20,23
15	230	Ecological indicators: Between the two fires of science and policy	Turnhout E., Hisschemöller M., Eijsackers H	2007	19,17
16	222	Urban Transition Labs: Co-creating transformative action for sustainable cities	Nevens F., Frantzeskaki N., Gorissen L., Loorbach D	2013	37,00
17	219	Chemisorption on metal surfaces	Norsko J. K	1990	7,55

(continued)

Table 2 (continued)

R	TC	Title	Author/s	Year	C/Y
18	203	Recent experimental progress in nuclear halo structure studies	Tanihata I., Savajols H., et al	2013	33,83
19	203	Selection of rational dispute resolution method by applying new step-wise weight assessment ratio analysis (SWARA)	Keršuliene V., Zavadskas E. K., Turskis Z	2010	22,56
20	200	Size-consistent wave functions for nondynamical correlation energy: The valence active space optimized orbital coupled-cluster doubles model	Krylov A. I., Sherrill C. D., Byrd E. F. C., Head-Gordon M	1998	9,52
21	187	Hardware-efficient variational quantum eigensolver for small molecules and quantum magnets	Kandala A., Mezzacapo A., Temme K., Takita M., Brink M., Chow J. M., Gambetta J. M	2017	93,50
22	187	Finding density functionals with machine learning	Snyder J. C., Rupp M., Hansen K., Müller K.-R., Burke K	2012	26,71
23	184	Locally self-consistent Green's function approach to the electronic structure problem	Abrikosov I., Simak S., Johansson B., Ruban A., Skriver H	1997	8,36
24	181	Facilitated modelling in operational research	Franco L. A., Montibeller G	2010	20,11
25	176	The inverse band-structure problem of finding an atomic configuration with given electronic properties	Franceschetti A., Zunger A	1999	8,80
26	176	Coping with intractable controversies: The case for problem structuring in policy design and analysis	Hisschemöller M., Hoppe R	1995	7,33
27	175	Order-N multiple scattering approach to electronic structure calculations	Wang Y., Stocks G. M., Shelton W. A., Nicholson D. M. C., Szotek Z., Temmerman W. M	1995	7,29
28	169	Effective dielectric constant of periodic composite structures	Datta S., Chan C. T., Ho K. M., Soukoulis C. M	1993 6,50	
29	158	Subsystem density-functional Theory	Jacob C. R., Neugebauer J	2014	31,60
30	158	Cognitive mapping expert views for policy analysis in the public sector	Eden C., Ackermann F	2004	10,53
31	158	Aporia: A critical exploration of the agent-structure problematique in international relations theory	Doty R. L	1997	7,18
32	146	Potential scattering, transfer matrix, and group theory	Alhassid Y., Gürsey F., Iachello F	1983	4,06

(continued)

Table 2 (continued)

R	TC	Title	Author/s	Year	C/Y
33	137	An analysis of the Six Sigma DMAIC method from the perspective of problem solving	De Mast J., Lokkerbol J	2012	19,57
34	133	Problem structuring using computer-aided morphological analysis	Ritchey T	2006	10,23
35	133	Conservative load projection and tracking for fluid–structure problems	Cebral J. R., Löhner R	1997	6,05

Abbreviations R = Rank
TC = Total citations
C/Y = Citations per year

Table 5 presents the thirty most productive countries in Problem Structuring. USA is already the most productive and influential country on the topic, although still very close is to the United Kingdom. Under this perspective, China and Germany obtain the most remarkable results being in third and fourth place respectively, in absolute terms.

3.3 Graphical Analysis of PS with VOSviewer

The previous section provides a general description of the most productive and influential variables in the journal. Therefore, to provide a deeper analysis of the data obtained from the search, let us visualize the bibliographic connections between the leading sources. For doing so, the work uses the VOS viewer. This program can be used to build and view graphic maps through the use of bibliographic coupling [52]. Likewise, analysis of citations and co-citations, co-authorship and cooccurrence of author keywords [30, 34, 35].

First, let us analyze co-citation that occurs when two documents published in different journals receive a citation from the third document of another journal [35]. (See Fig. 2) presents the general visualization between 1980 and 2019 with a minimum threshold of twenty cites and the two hundred most representative connections. Note that the colors of the circles of the journals indicate the cluster to which the journals belong.

Next, let us analyze the bibliographic coupling of authors that publish PS (see Fig. 3) presents the most productive authors with a minimum threshold of three published documents and the thirty-five strongest bibliographic coupling links between authors.

The advantage of this figure is the graphical mapping of authors that connects or clusters together those that have similar research profiles. That is, those that cites

Table 3 The most productive authors in PS

R	Author Name	University	Country	TP	TC	H	C/P	≥ 200	≥ 100	≥ 50
1	White L	The University of Warwick	United Kingdom	12	294	21	24.50	0	0	2
2	Ackermann F	Curtin University, School of Management	Australia	11	740	33	67.27	0	2	0
3	Rosenhead J	London School of Economics and Political Science	United Kingdom	11	719	17	65.36	1	1	2
4	Shaw D	Nottingham University Business School	United Kingdom	11	253	20	23.00	0	0	0
5	Yearworth M	University of Exeter	United Kingdom	11	98	10	8.91	0	0	0
6	Franco L. A	Universidad del Pacífico	Peru	10	453	20	45.30	0	1	2
7	Belton V	Strathclyde Business School	United Kingdom	9	500	24	55.56	0	1	4
8	Paucar-Caceres A	University of Hull	United Kingdom	9	42	12	4.67	0	0	0
9	Midgley G	Manchester Metropolitan University	United Kingdom	8	241	23	30.13	0	0	1
10	Eden C	Open University	United Kingdom	7	719	37	102.71	1	1	2
11	Bell S	Massachusetts Institute of Technology	United States	7	93	16	13.29	0	0	0
12	Gnesin V	University of Strathclyde	United Kingdom	7	52	6	7.43	0	0	0
13	Mingers J	University of Texas at Austin	United States	6	605	38	100.83	1	0	3
14	Montibeller G	Dalhousie University	Canada	6	389	6	64.83	0	1	2
15	Chelikowsky J. R	Max Planck Institute for Solid State Research	Germany	6	76	71	12.67	0	0	0
16	Sigrist J.-F	DCNS Research	France	6	49	12	8.17	0	0	0
17	Chen Z	Panepistimion Pireos	Greece	6	47	27	7.83	0	0	0
18	Assimakopoulos N. A	National Academy of Sciences	Ukraine	6	18	6	3.00	0	0	0

(continued)

Table 3 (continued)

R	Author Name	University	Country	TP	TC	H	C/P	≥ 200	≥ 100	≥ 50
19	Burke K	University of California, Irvine,	United States	5	887	52	177.40	1	2	0
20	Antunes C. H	Universidade de Coimbra	Portugal	5	110	34	22.00	0	0	1
21	Humphreys P	London School of Economics and Political Sciencea	United Kingdom	5	103	7	20.60	0	0	1
22	Tavella E	Københavns Universitet, Department of Food and Resource Economic	Denmark	5	37	6	7.40	0	0	0
23	Belderrain M. C. N	Instituto Tecnologico de Aeronautica,	Brazil	5	14	5	2.80	0	0	0
24	Scuseria G. E	Rice University, Department of Chemistry	United States	4	199	108	49.75	0	0	1
25	Li G	Shenyang Institute oEcology Chinese Academy of Sciences	China	4	170	40	42.50	0	1	1
26	Hämäläinen R.P	University, Systems Analysis Laboratory	Finland	4	158	32	39.50	0	0	1
27	Kübler J	Technische Universität Darmstadt	Germany	4	137	47	34.25	0	1	0
28	Farhat C	Stanford University	United States	4	129	66	32.25	0	0	1
29	Elliott M	University of Hull	United Kingdom	4	109	52	27.25	0	0	1
30	Giordano R	University of Minnesota	Italy	4	90	14	22.50	0	0	1
31	Belytschko T	Northwestern University, Department of Mechanical Engineering, Evanston	United States	4	88	117	22.00	0	0	1
32	Dias L. C	Universidade de Coimbra,	Portugal	4	73	24	18.25	0	0	1
33	Rzadkowski R	Institute of Fluid-Flow Machinery of the Polish Academy of Sciences, Gdansk, Poland	Poland	4	52	11	13.00	0	0	0
34	Saad Y	University of Minnesota	United States	4	56	52	14.00	0	0	0

(continued)

Table 3 (continued)

R	Author Name	University	Country	TP	TC	H	C/P	≥ 200	≥ 100	≥ 50
35	Papadopoulos T	University of Kent	United Kingdom	4	46	31	11.50	0	0	0

Abbreviations available in except: C/P = Cites per paper; H = h-index

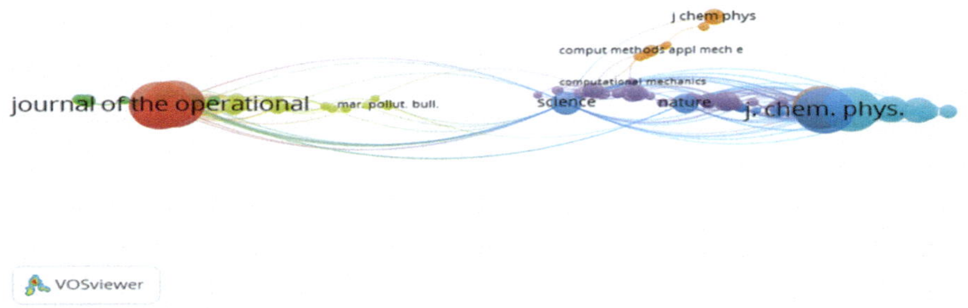

Fig. 2 Co-citation of journal cited in PS

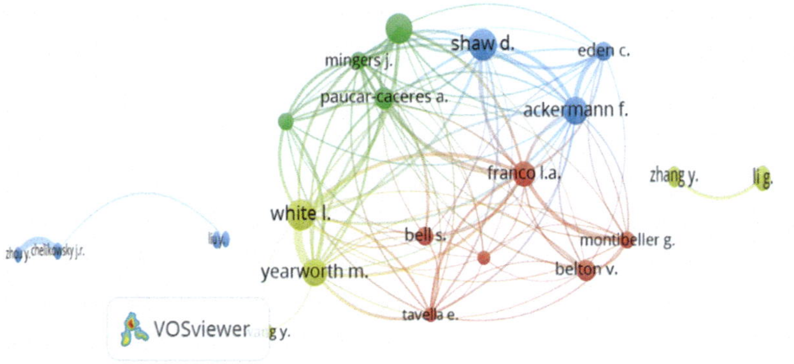

Fig. 3 Bibliographic coupling of authors that publish in PS

similar bibliographic material. It is worth mentioning that the results of this figure are in accordance with Table 3.

Next, let us look into the bibliographic coupling of institutions publishing in the journal (see Fig. 4) visualizes the data considering a minimum threshold of two documents and thirty-five connections.

The leading institutions are consistent with those in Table 4. We observed that the bibliographic data of the institutions depend on the authors who work in these organizations. To deepen the analysis, we presented a graphic visualization of the appointments between institutions. To do this, (see Fig. 5) shows the results with the same threshold and connections as (see Fig. 4). In addition, it can be seen that Victoria University of Welling is not the one with the most documents and citations registered, but has a strong relationship with several institutions. The University of Warwick has the most citations, but has a smaller number of connections with the universities.

Institutional co-authorship (see Fig. 6) presents the same threshold and connections as la (see Figs. 4 and 5). The results are similar to those of the previous figures.

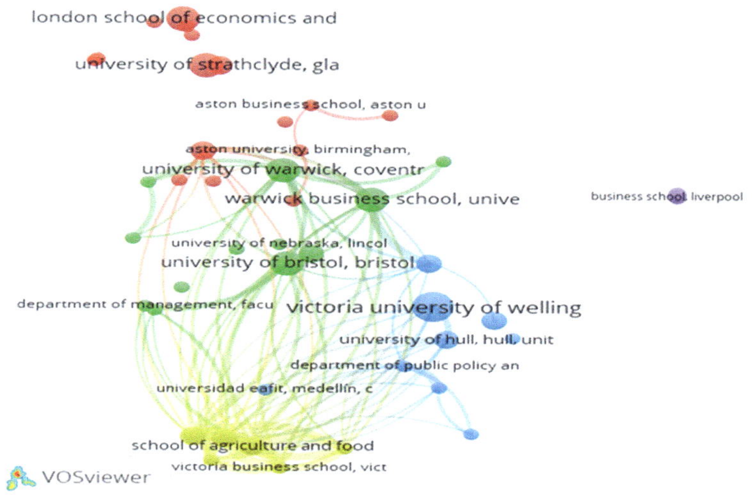

Fig. 4 Bibliographic coupling of institutions that publish in PS

In order to summarize the results at the country level, (see Fig. 7) presents bibliographic coupling of countries with a threshold of one document and one hundred-twelve connections. The results are in accordance with Table 5 where the USA is the most productive country followed by the United Kingdom. It is observed that the countries in the first positions are related to most other countries.

(See Fig. 8) visualizes the map considering a threshold of five occurrences and the thirty-five most representative connections. It is noted that the larger node refers to the keyword, Problem Structuring, as a central theme that has been carried out for the development of the research field. The above, followed by the nodes of Problem structuring methods, soft or, facilitation, decision analysis, soft systems methodology, System dynamics, Cognitive mapping, Process, and methodology are the most frequent keywords in the magazine.

4 Conclusions

Problem structuring is a topic that continues being of interest in the academic community. It was showed in this bibliometric study, which included 1550 articles and reviews since 1980 to 2019 from Scopus database. The data gathering process was conducted by using the VOS Viewer computer program. Previously, a set of criteria were defined as bibliometric indicators. With collected data was possible to do a mapping analysis about problem structuring topic. This mapping visually shows the bibliometric networks affording with the indicator such as authors, articles, universities, countries, keywords, among others.

Table 4 Top 35 the most productive and influential institutions in PS

R	Institution	Country	TP	TC	C/P	≥ 200	≥ 100	≥ 50	ARWU	QS
1	The University of Warwick	United Kingdom 38 995			26.18	1	1	3	101–150 62	
2	London School of Economic and Political Science	United Kingdom	29	1384	47.72	1	2	6	151–200	44
3	University of Strathclyde	United Kingdom	23	1579	68.65	1	3	7	601–700	296
4	University of Hull	United Kingdom	21	553	26.33	0	0	3	901–1000	601–650
5	Russian Academy of Sciences	Russian	17	115	6.76	0	0	1	-	-
6	CNRS Centre National de la Recherche Scientifique	France	16	224	14.00	0	0	1	-	-
7	Victoria University of Wellington	New Zealand	15	307	20.47	0	0	1	301–400	
8	University of Texas at Austin	United States	15	309	20.60	0	0	1	45	63
11	University of Bristol	United Kingdom	14	279	19.93	0	0	2	64	49
12	University of Cambridge	United Kingdom	14	186	13.29	0	0	1	3	7
13	Aston University	United Kingdom	13	305	23.46	0	0	0	-	442

(continued)

Table 4 (continued)

R	Institution	Country	TP	TC	C/P	≥ 200	≥ 100	≥ 50	ARWU	QS
14	University of Illinois at Urbana-Champaign	United States	12	1267	105.58	2	2	0	38	71
15	Tsinghua University	China	12	327	27.25	0	1	2	43	16
16	ETH Zürich swiss federal institute technology	Switzerland	12	390	32.50	1	0	0	19	6
17	Delft University of Technology	Netherlands	11	162	14.73	0	0	1	151–200	52
18	University of Manchester	United Kingdom	10	156	15.60	0	0	0	33	27
19	Arizona State University	United States	10	320	32.00	0	1	1	101–150	212
20	University of California, Irvine	United States	10	1149	114.90	2	2	1	80	169
21	Chinese Academy of Sciences	China	10	173	17.30	0	0	0	–	–
22	Oak Ridge National Laboratory	United States	10	250	25.00	0	1	0	–	–
23	National Academy of Sciences in Ukraine	Ukraine	9	74	8.22	0	0	0	-	-
24	Polish Academy of Sciences	Poland	9	56	6.22	0	0	0	–	–
25	Shanghai Jiao Tong University	China	9	37	4.11	0	0	0	82	60

(continued)

Table 4 (continued)

R	Institution	Country	TP	TC	C/P	≥ 200	≥ 100	≥ 50	ARWU	QS
26	University of Queensland	Australia	9	234	26.00	0	0	1	54	47
27	Nanjing University	China	9	45	5.00	0	0	0	151–200	120
28	University of Piraeus	Greece	8	55	6.88	0	0	0	–	–
29	Technical University of Berlin	Germany	8	423	52.88	1	1	0	301–400	147
30	Open University	United Kingdom	8	143	17.88	0	0	0	701–800	601
31	National University of Singapore	Singapore	8	70	8.75	0	0	0	67	11
32	University of Twente	Netherlands	8	128	16.00	0	0	0	401–500	172
33	Manchester Metropolitan University	United Kingdom	8	43	5.38	0	0	0	801–900	751–800
34	UCL University College London	United Kingdom	8	188	23.50	0	0	0	15	8
35	University of Minnesota	United States	8	141	17.63	0	0	1	41	156

Abbreviations
Abbreviations are in except: ARWU and QS = Ranking in the general ARWU and QS university rankings

Table 5 Top 30 the most productive and influential countries in PS

R	Country	TP	TC	H	C/P	≥200	≥100	≥50	P/Po	C/Po
1	USA	370	12,090	55	32.68	9	14	36	1,12	36,74
2	United Kingdom	258	7873	47	30.52	3	7	32	3,87	118,12
3	China	152	1575	19	10.36	1	2	3	0,10	1,09
4	Germany	88	3063	22	34.81	4	4	4	1,05	36,67
5	France	84	1831	21	21.80	2	0	4	1,28	28,11
8	Canada	65	1676	21	25.78	2	2	5	1,73	44,80
7	Netherlands	62	1879	21	30.31	2	4	3	3,62	109,88
9	Italy	58	1686	19	29.07	1	2	6	0,95	27,85
10	Brazil	49	642	13	13.10	0	0	4	0,23	3,04
11	Australia	48	724	16	15.08	0	0	2	1,90	28,72
12	Japan	42	789	13	18.79	1	1	2	0,33	6,21
13	Russian	38	694	8	18.26	1	0	1	0,26	4,75
14	Switzerland	33	795	15	24.09	1	0	3	3,84	92,54
15	Spain	29	651	14	22.45	0	1	4	0,62	13,92
16	New Zealand	25	517	15	20.68	0	1	0	5,23	108,15
17	Portugal	25	815	12	32.60	1	0	1	2,44	79,66
18	Denmark	24	1057	15	44.04	1	3	1	4,15	183,18
19	Korea South	24	668	8	27.83	1	2	1	0,46	13,03
20	Finland	21	447	10	21.29	0	0	2	3,79	80,83
21	India	21	165	9	7.86	0	0	0	0,01	0,12
22	Sweden	21	657	12	31.29	0	2	1	2,09	65,43
23	Poland	20	132	5	6.60	0	0	0	0,52	3,48
24	Austria	19	321	11	16.89	0	0	0	2,12	35,82
25	Belgium	18	589	9	32.72	1	0	2	1,55	51,03
26	Iran	17	133	7	7.82	0	0	0	0,20	1,60
27	Turkey	17	302	8	17.76	0	0	1	0,20	3,61
28	Greece	16	194	7	12.13	0	0	1	1,52	18,52
29	Singapore	15	253	6	16.87	0	1	1	2,58	43,62
30	Ukraine	14	106	5	7.57	0	0	0	0,31	2,40
31	Israel	12	434	8	36.17	0	2	0	1,40	50,93
32	Norway	12	102	5	8.50	0	0	0	2,23	18,95
33	Taiwan	12	407	8	33.92	0	1	2	0,50	17,22
34	Chile	10	118	5	11.80	0	0	0	0,52	6,22
35	Hong Kong	10	149	6	14.9	0	0	1	1,34	20,03

Abbreviations available in previous tables except: P/Po and C/Po = Papers and cites per million inhabitants

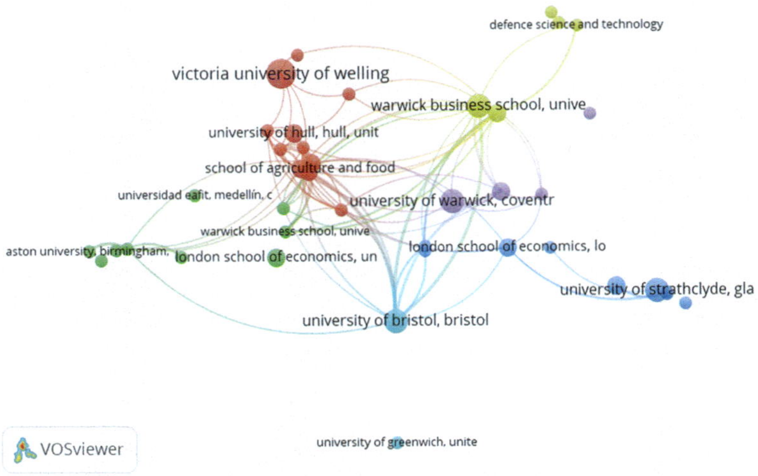

Fig. 5 Citation analysis of institutions publishing in PS

Fig. 6 Co-authorship of institutions publishing in PS

According to the results, it is possible to identify that White L. from University of Warwick is the most important author in problem structuring. This university occupies the first position in the top 35 of the most productive and influential institutions about Problem Structuring. The University of Warwick is in the United Kingdom, the country with the second most productive and influential place. Likewise, United Kingdom has thirteen universities within the top 35. It represents 37% of the total of the institutions in the top 35; where the USA is home of 17.14% of institutions, but it is in first place of the ranking.

In the same sense, the most cited article related to problem structuring is "The Agent-Structure Problem in International Relations Theory" of Wendt A. E 1987. There is no relationship with White L., the most productive author on this topic. It is interesting the most common keyword occurrence was "problem structuring" per se, with 128 occurrences.

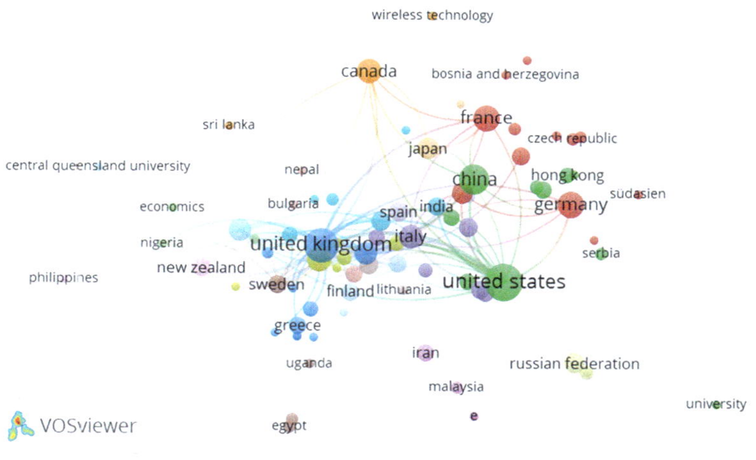

Fig. 7 Bibliographic coupling of countries that publish in PS

Fig. 8 Co-occurrence of author keywords of documents published in PS

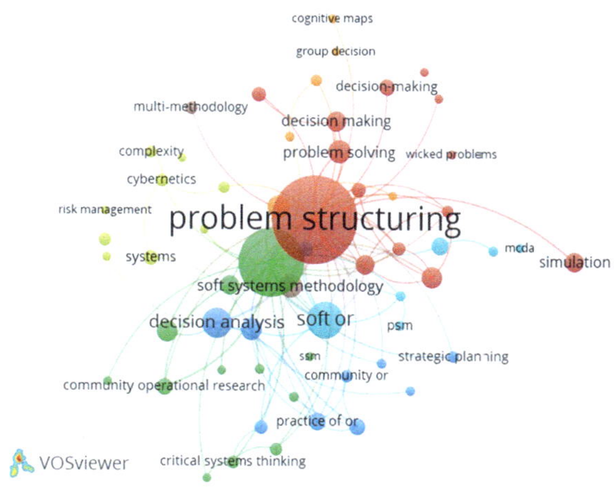

As a final summary, the main authors in problem structuring are White L., Ackermann F., and Rosenhead J. The main institutions are University of Warwick, London School, and University of Strathclyde, located in the United Kingdom. Therefore, the main productive countries in problem structuring studies are the USA, the United Kingdom, and China, and the main citation countries are USA, the United Kingdom, and Germany. The findings provide a pathway to practitioners of problem structuring to build a deeply theoretical framework on the subject, and an appropriate analysis of new developments in the area. Authors hope this paper can be useful to support the researching of problem structuring.

References

1. Abrikosov I, Simak S, Johansson B, Ruban A, Skriver H (1997) Locally self-consistent Green's function approach to the electronic structure problem. Phys Rev B-Condens Matter Mater Phys 56(15):9319–9334
2. Ackermann F (2012) Problem structuring methods ' in the Dock ': Arguing the case for Soft OR. Eur J Oper Res 219(3):652–658
3. Ackermann F (2015) Teaching soft OR/MS methodologies: what, why, and how. Wiley Encycl Oper Res Manag Sci 1–9
4. Ackermann F, Franco LA, Rouwette E, White L (2014) Special issue on problem structuring research and practice. EURO J Decis Process 2(3–4):165–172
5. Alhassid Y, Gürsey F, Iachello F (1983) Potential scattering, transfer matrix, and group theory. Phys Rev Lett 50(12):873–876
6. Atalla N, Panneton R, Debergue P (1998) A mixed displacement-pressure formulation for poroelastic materials. J Acoust Soc Am 104(3):1444–1452
7. Belton V, Stewart T (2010) Problem structuring and MCDA. In: Ehrgott M, Figueira J (eds) International series in operations research and management science, trends in multiple criteria decision analysis vol 142. Springer, pp 209–239
8. Ben-Nun M, Quenneville J, Martínez TJ (2000) Ab initio multiple spawning: photochemistry from first principles quantum molecular dynamics. J Phys Chem A 104(22):5172–5175
9. Burke K (2012) Perspective on density functional theory. JChem Phys 136(15)
10. Blanco-Mesa F, Merigó JM, Gil-Lafuente AM (2017) Fuzzy decision making: a bibliometric-based review. J Intell Fuzzy Syst 32(3):2033–2050
11. Blanco-Mesa F, León-Castro E, Merigó JM (2019) A bibliometric analysis of aggregation operators. Appl Soft Comput J 81:105488
12. Caputo A, Marzi G, Pellegrini MM, Rialti R (2018) Conflict management in family businesses: a bibliometric analysis and systematic literature review. Int J Confl Manag 29(4):519–542
13. Causin P, Gerbeau JF, Nobile F (2005) Added-mass effect in the design of partitioned algorithms for fluid-structure problems. Comput Methods Appl Mech Eng 194(42–44):4506–4527
14. Cebral JH, Löhner R (1996) Conservative load projection and tracking for fluid-structure problems. 34th Aerosp Sci Meet Exhib 35(4):687–692
15. De Mast J, Lokkerbol J (2012) An analysis of the Six Sigma DMAIC method from the perspective of problem solving. Int J Prod Econ 139(2):604–614
16. Dias R, Cabral AS, López B, Belderrain MCN (2016) The use of cognitive maps for requirements elicitation in product development. J Aerosp Technol Manag 8(2):178–192. https://doi.org/10.5028/jatm.v8i2.578
17. Dick AS, Basu K (1987) From the SAGE social science collections. All Rights. Hisp J Behav Sci 9(2):183–205
18. Eden C (2004) Analyzing cognitive maps to help structure issues or problems. Eur J Oper Res 159(3):673–686
19. Eden C, Ackermann F (2004) Cognitive mapping expert views for policy analysis in the public sector. Eur J Oper Res 152(3):615–630
20. Franceschetti A, Zunger A (1999) The inverse band-structure problem of finding an atomic configuration with given electronic properties. Nature 402(6757):60–63
21. Franco LA, Montibeller G (2010) Facilitated modelling in operational research. Eur J Oper Res 205(3):487–734
22. Franco LA, Lord E (2011) Understanding multi-methodology: evaluating the perceived impact of mixing methods for group budgetary decisions. Omega 39(3):362–372
23. Hisschemöller M, Hoppe R (1996) Coping with interactable controversies: The case for problem structuring in policy design and analysis. Knowl Policy: Int J Knowl Transf Util 8(4):40–60
24. Jackson K, Pederson MR (1990) Accurate forces in a local-orbital approach to the local-density approximation. Phys Rev B 42(6):3276–3281

25. Jacob CR, Neugebauer J (2014) Subsystem density-functional theory. Wiley Interdiscip Rev: Comput Mol Sci 4(4):325–362
26. Kandala A, Mezzacapo A, Temme K, Takita M, Brink M, Chow JM, Gambetta JM (2017) Hardware-efficient variational quantum eigensolver for small molecules and quantum magnets. Nature 549(7671):242–246
27. Kessler MM (1963) Bibliographic coupling between scientific papers. Am Doc 14(1):10–25
28. Krylov AI, Sherrill CD, Byrd EFC, Head-Gordon M (1998) Size-consistent wave functions for nondynamical correlation energy: the valence active space optimized orbital coupled-cluster doubles model. J Chem Phys 109(24):10669–10678
29. Lichtenstein A, Katsnelson (1998) Ab initio calculations of quasiparticle band structure in correlated systems: LDA++ approach. Phys RevB Condens Matter Mater Phys 57(12):6884–6895
30. Martínez-López FJ, Merigó JM, Gázquez-Abad JC, Ruiz-Real JL (2020) Industrial marketing management: Bibliometric overview since its foundation. Indus Mark Manag 84(July):19–38
31. Marx D (2006) Proton transfer 200 years after Von Grotthuss: insights from ab initio simulations. Chem Phys Chem 7(9)
32. Mendoza GA, Martins H (2006) Multi-criteria decision analysis in natural resource management: a critical review of methods and new modelling paradigms. For Ecol Manage 230(1–3):1–22
33. Merigó JM, Mas-Tur A, Roig-Tierno N, Ribeiro-Soriano D (2015) A bibliometric overview of the journal of business research between 1973 and 2014. J Bus Res 68(12):2645–2653
34. Merigó JM, Cancino CA, Coronado F, Urbano D (2016) Academic research in innovation: a country analysis. Scientometrics 108(2):559–593
35. Merigó JM, Yang JB (2017) A bibliometric analysis of operations researches and management science. Omega-Int J Manag Sci 73:37–48
36. Mingers J, Rosenhead J (2004) Problem structuring methods in action. Eur J Oper Res 152(3):530–554
37. Moed HF (2009) New developments in the use of citation analysis in research evaluation. Arch Immunol Ther Exp 57(1):13–18
38. Nevens F, Frantzeskaki N, Gorissen L, Loorbach D (2013) Urban transition labs: co-creating transformative action for sustainable cities. J Clean Prod 50:111–122
39. Norsko JK (1990) Chemisorption on metal surfaces. Rep Prog Phys 53(10):1253–1295
40. Paul R, Tuszyński JA, Chatterjee R (1984) Dielectric constant of biological systems. Phys Rev A 30(5):2676–2685
41. Ritchey T (2006) Problem structuring using computer-aided morphological analysis. J Oper Res Soc 57(7):792–801
42. Rosenhead J (2006) Past, present and future of problem structuring methods. J Oper Res Soc 57(7):759–765
43. Small H (1973) Co-citation in the relationship between two documents. J Am Soc Inf Sci 24:256–269
44. Small H (2010) Emerging research fronts in science and technology: patterns of new knowledge development. Scientometrics 83:15–38
45. Snyder JC, Rupp M, Hansen K, MÜller KR, Burke K (2012) Finding density functionals with machine learning. Phys Rev Lett 108(25):1–5
46. Storck S, Bretinger H, Maier WF (1998) Characterization of micro- and mesoporous solids by physisorption methods and pore-size analysis. Appl Catal A 174(1–2):137–146
47. Tanihata I, Savajols H, Kanungo R (2013) Recent experimental progress in nuclear halo structure studies. Prog Part Nucl Phys 68(1):215–313
48. Teter MP, Payne MC, Allan DC (1989) Solution of Schrödingers equation for large systems. Phys Rev B 40(18):12255–12263
49. Turnhout E, Hisschemöller M, Eijsackers H (2007) Ecological indicators: between the two fires of science and policy. Ecol Ind 7(2):215–228
50. Valenzuela-Fernandez L, Merigó JM, Lichtenthal JD, Nicolas C (2019) A bibliometric analysis of the first 25 years of the journal of business-to-business marketing. J Bus Bus Mark 26(1):75–94

51. Valenzuela LM, Merigó JM, Johnston WJ, Nicolas C, Jaramillo JF (2017) Thirty years of the Journal of Business and Industrial Marketing: a bibliometric analysis. J Bus Indus Mark 32(1):1–17
52. Van Eck NJ, Waltman L (2010) Software survey: VOSviewer, a computer program bibliometric mapping. Scientometrics 84(2):523–538
53. Wang Y, Stocks M, Shelton A, Szotek Z, Recently N (1995) Order-N multiple scattering approach to electronic struture calculations. Phys Rev Lett 75(15):2867–2870
54. Wendt AE (1987) The agent-structure problem in international relations theory. Int Organ 41(3):335–370
55. White L (2006) Evaluating problem-structuring methods: developing an approach to show the value and effectiveness of PSMs. J Oper Res Soc 57(7):842–855
56. White L, Burger K, Yearworth M (2016) Understanding behaviour in problem structuring methods interventions with activity theory. Eur J Oper Res 249(3):983–1004
57. Woolley RN, Pidd M (1981) Problem structuring— a literature review. J Oper Res Soc 32(32):197–206

Chapter 16
Study of Some Properties and Behavior of the Group of Signed Quadratic Residues and Its Applications to Cryptography

Pinkimani Goswami and Madan Mohan Singh

1 Introduction

For an odd positive integer N, let us denote QR_N as the group of quadratic residues modulo N. If the set $\mathbb{Z}_N$ is represented as $\left\{\frac{-(N-1)}{2}, \ldots, -1, 0, 1, \ldots, \frac{N-1}{2}\right\}$ then $|\xi|, \xi \in \mathbb{Z}_N$, denotes the absolute value of ξ. The set $QR_N^+ = \{|\xi| : \xi \in QR_N\}$ forms a group under the operation "$\circ$" which is defined as follows: $\lambda \circ \xi = |\lambda \cdot \xi (\mathrm{mod} N)|, \forall \lambda, \xi \in QR_N^+$. The group QR_N^+ is called the group of signed quadratic residue modulo N. In 2009 [1], Hofheinz et al. have showed some advantages of this group in cryptography. The group QR_N^+ has satisfied various intractable properties like a quadratic residue group. The main advantage of this group is that the elements of this group are easily observable and therefore it is useful to construct an encryption scheme having IND-CCA (Indistinguishability-Chosen Ciphertext Attack) security. This is because a decryption oracle can determine whether a given ciphertext is signed quadratic residue or not and hence can debar an attacker for availing any secret information. Note that, QR_N^+ is an example of Gap group [2], in which decisional problem is easy but the corresponding computational problem is hard. Despite their various interesting properties, they have found very little cryptographic applications so far. Little bit of applications can be found in papers [1, 3–14] only. In this paper, we bring out some of those applications together for motivating researcher.

P. Goswami (✉)
University of Science and Technology Meghalaya, Techno City, India
e-mail: pinkimanigoswami@yahoo.com

M. M. Singh
North-Eastern Hill University, Shillong, Meghalaya, India

M. Sahni et al. (eds.), *Novel Developments in Futuristic AI-based Technologies*, Algorithms for Intelligent Systems, https://doi.org/10.1007/978-981-99-3076-0_16

In this paper, we will review the properties of signed quadratic residues and describe the techniques for implementation of signed quadratic residue group to cryptography. The subsequent part of the paper is organized as follows: In Sect. 2, we discussed about the merits and demerits of quadratic residue group for implementing cryptography and the motivation behind the construction of signed quadratic residue group. Section 3 is devoted to the signed quadratic residue group. In Sect. 4, we discussed the applications of signed quadratic residue group in cryptography. Necessary conclusion has been made in Sect. 5.

2 Quadratic Residues

For any positive integer N, an integer ξ with $\gcd(\xi, N) = 1$ is said to be a quadratic residue modulo N if $x^2 \equiv \xi \bmod N$ for some integer x. In this case, x is called a square root of ξ modulo N. In absence of any such x, the integer ξ is a quadratic non-residue modulo N. The set QR_N stand for all quadratic residues modulo N and it is a subgroup of $\mathbb{Z}_N^*$, the multiplicative group modulo N. For $N = \Pi_{i=1}^{r} P_i^{k_i}$ $(P_1 < P_2 \cdots < P_r)$, the cardinality of QR_N is $\frac{\phi(N)}{2^r}$. In fact, QR_N is a cyclic group if and only if:

(a) For N is odd, $\gcd\left(\frac{\phi\left(P_i^{k_i}\right)}{2}, \frac{\phi\left(P_j^{k_j}\right)}{2}\right) = 1$, for all $1 \leq i < j \leq r$.

(b) For N is even,

- $\gcd\left(\frac{\phi\left(P_i^{k_i}\right)}{2}, \frac{\phi\left(P_j^{k_j}\right)}{2}\right) = 1$, for all $1 \leq i < j \leq r$ and
- if 2^k is the largest power of 2 that divide N and $k > 3$, then $P_i \equiv 3(\text{mod}4), \forall i > 1$.

According to Euler's criterion, for any odd prime P, $\xi \in QR_P$ iff $\xi^{\frac{P-1}{2}} \equiv 1(\text{mod}P)$, which implies that there is a polynomial time algorithm for checking $\xi \in QR_P$ or not [15]. Also, for $k > 1$, $\xi \in QR_{P^k} \iff \xi \in QR_P$ and hence one can easily check whether $\xi \in QR_{P^k}$ or not. Similarly, for $N = \Pi_{i=1}^{r} P_i^{k_i}$, $k_i \geq 1$, $\xi \in QR_N \iff \xi \in QR_{P_i}, \forall i$. Therefore, if the factorization of N is known, then one can easily check whether ξ is a quadratic residue or not. However, an efficient algorithm to check that $\xi \in QR_N$ or not if the factors of N are unknown is absent till date [15]. There are numerous encryption schemes whose security relies on this idea, one such example is RSA cryptosystem [16].

For the completeness of this section, let us define the following symbols:

Legendre Symbol: For an odd prime P and for any integer ξ, the Legendre symbol $\left(\frac{\xi}{P}\right)$ is defined as follows:

$$\left(\frac{\xi}{P}\right) = \begin{cases} -1 \text{ if } \xi \notin QR_P \\ 0 \text{ if } P|\xi \\ 1 \text{ if } \xi \in QR_P \end{cases}$$

From the definition, it is clear that one can efficiently compute $\left(\frac{\xi}{P}\right)$.

Jacobi Symbol: Suppose $N = \Pi_{i=1}^{r} P_i^{k_i}$ is an odd positive integer and let ξ be any integer. Then the Jacobi symbol $\left(\frac{\xi}{N}\right)$ is defined as follows:

$$\left(\frac{\xi}{N}\right) = \Pi_{i=1}^{r}\left(\frac{\xi}{P_i}\right)^{k_i}$$

where the notation $(\cdot)$ of R.H.S. denotes Legendre symbol. The following properties of the Jacobi symbol help one to efficiently compute $\left(\frac{\xi}{N}\right)$, without knowing the factors of N [15]:

(a) If $\lambda \equiv \xi \pmod{N}$, then $\left(\frac{\lambda}{N}\right) = \left(\frac{\xi}{N}\right)$.

(b) $\left(\frac{2}{N}\right) = \begin{cases} -1 \text{ if } N \equiv \pm 3 \bmod 8 \\ 1 \text{ if } N \equiv \pm 1 \bmod 8 \end{cases}$

(c) $\left(\frac{\lambda\xi}{N}\right) = \left(\frac{\lambda}{N}\right)\left(\frac{\xi}{N}\right)$

(d) For any odd positive integer K,

$$\left(\frac{K}{N}\right) = \begin{cases} -\left(\frac{N}{K}\right) \text{if } K \equiv N \equiv 3 \bmod 4 \\ \left(\frac{N}{K}\right) \text{otherwise} \end{cases}$$

Let $J(N) = \left\{\xi \in \mathbb{Z}_N^* : \left(\frac{\xi}{N}\right) = 1\right\}$. Note that, without factorization of N one can compute $\left(\frac{\xi}{N}\right)$ and if $\xi \notin J(N)$, then it guaranteed that$\xi \notin QR_N$, but if $\xi \in J(N)$, then this does not imply $\xi \in QR_N$. In fact, for given a composite number N and $\xi \in J(N)$ with $\gcd(\xi, N) = 1$, deciding whether $\xi \in QR_N$ or not, is called the Quadratic residuosity problem (QRP) [15]. It is assumed that the QRP is hard, and it is a stronger assumption than the factoring assumption [1]. There is a large number of cryptographic protocols whose security depends on this assumption (e.g., [17–20], etc.). Also, one can find the square roots modulo N if factors of N are known, but there is no efficient algorithm for computing square roots if factors of N are unknown. In fact, Rabin [21] has proved that computing square root modulo N is at least as hard as finding the factors of N. Some of the cryptosystems which are defined over quadratic residue group are found in [22–25].

Shmuely [26] and McCurley [27] have proved that in the group QR_N, the Diffie-Hellman (DH) problem is at least as hard as factoring N.

Despites its numerous interesting properties and cryptographic applications, it has some disadvantages too. Basically, a cryptosystem-based quadratic residuosity assumption may not achieve IND-CCA security, because the elements of QR_N is not efficiently distinguishable without the factorization of N. Therefore, an adversary may send both quadratic residue and quadratic non-residue to a decryption oracle, and can revile some secret information. Also, it is observed that, because of the indistinguishable property of QR_N, it is difficult to prove that the Strong Diffie-Hellman (SDH) problem over the group QR_N is at least as hard as factoring [1].

3 Fundamental of SQR

3.1 Signed Quadratic Residue Modulo N

Suppose N be an odd integer. The set $\mathbb{Z}_N = \{0, 1, \cdots N-1\}$ can also be represented as $\left\{-\frac{N-1}{2}, \cdots, -1, 0, 1, \cdots, \frac{N-1}{2}\right\}$. For any subgroup H of $\mathbb{Z}_N^*$, let us define a set H^+ of $\mathbb{Z}_N^*$ as follows:

$$H^+ := \{|\xi| : \xi \in H\}$$

where $|\xi|$ is denoted as the absolute value of ξ. Clearly, H^+ is a non-empty set and it forms a group w.r.t the operation "$\circ$", defined as follows: for $\lambda, \xi \in H^+$,

$$\lambda \circ \xi = |\lambda \cdot \xi (\text{mod} N)|.$$

For an integer n, define

$$\xi^{\underline{n}} := \underbrace{\xi \circ \xi \circ \ldots \circ \xi}_{n\, times} = \left|\xi^n (\text{mod } N)\right|.$$

Define a map $\psi : H \rightarrow H^+$ as $\psi(\xi) := |\xi|, \xi \in H$. Then

(1) The map ψ is a group homomorphism.

(2) $Ker\ \psi = \begin{cases} \{1\} & \text{if } -1 \notin H \\ \{1, -1\}, & \text{if } -1 \in H \end{cases}$

(3) If Ker $\psi = \{1\}$ then ψ is isomorphism.

In particular, the set QR_N is a subgroup of $\mathbb{Z}_N^*$ and hence one can construct the group QR_N^+. In 2009 [1], Hofheinz et al. named this group as the signed quadratic residues (in short SQR) group modulo N.

Some of the properties of QR_N^+ are discussed below. These results are considered from [1].

Lemma 1 *For any Blum integer* N(i.e., $N = PQ$, *where* $P \equiv Q \equiv 3 \pmod 4$ *are prime*),

(a) QR_N^+ *is a finite group with* $\frac{\phi(N)}{4}$ *elements.*
(b) *If* QR_N *is cyclic, then* QR_N^+ *is cyclic.*

(c) $QR_N^+ = J(N) \cap \left[\frac{N-1}{2}\right]$, where $\left[\frac{N-1}{2}\right] = \left\{1, 2, \cdots, \frac{N-1}{2}\right\}$

Proof Since the map $\psi : QR_N \to QR_N^+$ is a group homomorphism and Ker$\psi = \{1\}$, so ψ is a group isomorphism. Therefore, the cardinality of both the group is the same, i.e., $\frac{\phi(N)}{4}$. Again, for a Blum integer N, QR_N is a cyclic group and hence by the properties of group isomorphism, QR_N^+ is a cyclic group. Again, if $\xi \in QR_N^+$ then either $\xi \in QR_N$ or $-\xi \in QR_N$. If $\xi \in QR_N$ then $\left(\frac{\xi}{N}\right) = 1$. If $-\xi \in QR_N$ then $\left(\frac{-\xi}{N}\right) = 1$. But for a Blum integer N, $\left(\frac{\xi}{N}\right) = \left(\frac{-\xi}{N}\right) = 1$. Therefore, if $\xi \in QR_N^+$ then $\xi \in J(N) \cap \left[\frac{N-1}{2}\right]$. Conversely, if $\xi \in J(N) \cap \left[\frac{N-1}{2}\right]$, then either ξ or $-\xi$ in QR_N and therefore $\xi \in QR_N^+$. Hence $QR_N^+ = J(N) \cap \left[\frac{N-1}{2}\right]$.

Note that, from the above lemma, it is clear that the group QR_N^+, where N is a Blum integer, has achieved the same properties like quadratic residue group has, but in this group, one can easily distinguish its elements, without the knowledge of factors of N. Some other properties of QR_N^+ are as follows:

- Let N be a Blum integer and $\xi \in QR_N^+$. If $\gcd(\xi - 1, N) = 1 = \gcd(-\xi - 1, N)$, then ξ is a generator of QR_N^+ [9].
- Let N be a Blum integer. For an element $\xi \in QR_N^+$, the probability that $\xi \in QR_N$ is $\frac{1}{2} + o(1)$, where $o(1)$ is negligible [9].
- If $\lambda, \xi \in QR_N^+$, then $\lambda^2 \equiv \xi^2 \pmod N$ iff $\lambda \equiv \xi \pmod N$. Moreover, for an integer $k \geq 1$, $\lambda^{2^k} \equiv \xi^{2^k} \pmod N$ iff $\lambda \equiv \xi \pmod N$ [5].
- If $\lambda, \xi \in QR_N \cup QR_N^+$ then $|\lambda\xi| = ||\lambda||\xi||$ [5].

3.2 *Signed Quadratic Residues Modulo N^2*

The elements of $\mathbb{Z}_{N^2}$ can be represented uniquely as $\xi_1 + \xi_2 N$, where $\xi_1, \xi_2 \in \mathbb{Z}_N$. Also, the elements of the form $1 + \xi_1 N$ and $-1 + \xi_2 N$, where $\xi_1, \xi_2 \in \mathbb{Z}_N$ represents the elements of $\mathbb{Z}_{N^2}^*$ having order N and $2N$ respectively. For a Blum integer $N = PQ$, QR_{N^2} is a cyclic group and ord$QR_{N^2} = \frac{\phi(N^2)}{4} = \frac{N\lambda(N)}{2}$, with $\lambda(N) = \text{lcm}(P-1, Q-1)$, here $\lambda(N)$ is the Carmichael function for N. Note that, every element of order N are in QR_{N^2}. The signed quadratic residue group $QR_{N^2}^+$ is defined as

$$QR_{N^2}^+ = \{|\xi| : \xi \in QR_{N^2}\}$$

Since $-1 \notin QR_{N^2}$ so the function $\psi : QR_{N^2} \to QR_{N^2}^+$ is group isomorphism. Thus, the group $QR_{N^2}^+$ is also a cyclic group with cardinality $\frac{N\lambda(N)}{2}$. Also, $QR_{N^2}^+ = J(N^2) \cap \left[\frac{N^2-1}{2}\right]$, therefore the elements $QR_{N^2}^+$ are easily recognizable, where $J(N^2) = \left\{\xi \in \mathbb{Z}_{N^2}^* : \left(\frac{\xi}{N^2}\right) = 1\right\}$.

4 Cryptographic Applications of SQR

This section is devoted to deal with some applications of SQR in cryptography, which we have considered from [1, 3–10]. First, we will discuss some problems defined over QR_N^+ and $QR_{N^2}^+$.

4.1 RSA Problem and Its Variants Over QR_N^+

As it is proved that computing square roots modulo N is at least as hard as finding factors of N, therefore, finding the square roots over the group QR_N^+ (also $QR_{N^2}^+$) is hard, whereas one can easily determine the elements of this group. Therefore, QR_N^+ (also $QR_{N^2}^+$) is an example of Gap group. Note that, in a Gap group [2], a decisional problem is easy but the corresponding computational problem is hard.

RSA Problem [16]: Given a Blum integer N, an integer e with $\gcd(e, \phi(N)) = 1$ and a random element $\xi \in \mathbb{Z}_N^*$, RSA problem is to find $\lambda \in \mathbb{Z}_N^*$ such that $\lambda^e \equiv \xi \pmod{N}$.

Strong RSA Problem [28]: Given a Blum integer N, and a random element $\xi \in \mathbb{Z}_N^*$, strong RSA problem is to find an integer $e > 1$ and $\lambda \in \mathbb{Z}_N^*$ such that $\lambda^e \equiv \xi \pmod{N}$.

Strong QR-RSA Problem [29]: Given a Blum integer N, and a random quadratic residue $\xi \in QR_N$, Strong QR-RSA problem is to find an integer $e > 1$ and $\lambda \in \mathbb{Z}_N^*$ such that $\lambda^e \equiv \xi \pmod{N}$.

Strong SQR-RSA Problem [9]: Given a Blum integer N, and a random quadratic residue $\xi \in QR_N^+$, strong SQR-RSA problem is to find an integer $e > 1$ and $\lambda \in \mathbb{Z}_N^*$ such that $\lambda^e \equiv \xi \pmod{N}$.

In [9], it is proved that Strong SQR-RSA problem is as hard as Strong QR-RSA problem.

4.2 DH Problem and Its Variants Over QR_N^+

Consider $G =< \xi >$ be any cyclic group. For given ξ, ξ^a, and ξ^b in G, where a and b are unknown, the (computational) Diffie-Hellman (DH) Problem in G is to compute ξ^{ab}. In particular, the DH problem and the corresponding decisional problem over QR_N^+ are defined as follows:

DH Problem [1]: For given $N, \xi, X, Y \in QR_N^+$, where $X = \xi^a, Y = \xi^b$ with $a, b \in [1, \mathrm{ord} QR_N^+]$, the Diffie-Hellman problem is to compute $Z = \xi^{ab}$.

DDH Problem [1]: For given $N, \xi, X, Y \in QR_{N^2}^+$, where $X = \xi^a, Y = \xi^b$ with $a, b \in [1, \mathrm{ord} QR_N^+]$, the Decisional Diffie-Hellman (DDH) problem is to separate $Z = \xi^{ab}$ from a random group element in QR_N^+.

It is observed that as the element of QR_N^+ is efficiently recognizable therefore DDH problem over QR_N^+ is easy but DH problem is as hard as factoring N [1].

The *Strong Diffie-Hellman* (SDH) problem over a cyclic group G is to compute ξ^{ab}, for ξ, ξ^a, and ξ^b, in presence of a DDH oracle which returns 1 on (ξ^{ab}, ξ^c) if $\xi^{ab} = \xi^c$ and $(\xi^{ab}, \xi^c) \in G \times G$ or otherwise returns 0 [1]. In [1], Hofheinz et al. proved that SDH problem over QR_N^+ is hard, in spite of the presence of DDH oracle. They showed that without the factorization of N, it is hard to solve SDH problem over QR_N^+.

4.3 PDL Problem and Its Variants Over $QR_{N^2}^+$

In 1999 [30], Paillier defined the Partial Discrete Logarithm (PDL) Problem over a group G. The PDLP and some of its variants over $QR_{N^2}^+$ are defined as follows:

PDL Problem [7]: Let $\xi \in QR_{N^2}^+$ be an element with maximal order. For given ξ and$h = \xi^{\underline{a}} = |\xi^a (\mathrm{mod} N)|$, where $a \in [1, \mathrm{ord} QR_{N^2}^+]$, the PDL problem over $QR_{N^2}^+$ is to find $a(\mathrm{mod} N)$.

DDH Problem [7]: For given $N, \xi, X, Y \in QR_{N^2}^+$, where $X = \xi^{\underline{a}}, Y = \xi^{\underline{b}}$ with $a, b \in [1, \mathrm{ord} QR_{N^2}^+]$, the Decisional Diffie-Hellman (DDH) problem is to differentiate $Z = \xi^{\underline{ab}}$ from an arbitrary group element in $QR_{N^2}^+$.

S-CDH Problem [7]: For given $N, \xi, X, Y \in QR_{N^2}^+$, where $X = \xi^{\underline{a}}, Y = \xi^{\underline{b}}$ with $a \in \mathbb{Z}_N$ and $b \in [1, \mathrm{ord} QR_{N^2}^+]$, the Small Computational Diffie-Hellman (S-CDH) problem asks to compute $Z = g^{\underline{ab}}$.

S-DDH Problem [7]: For given $N, \xi, X, Y \in QR_{N^2}^+$, where $X = \xi^{\underline{a}}, Y = \xi^{\underline{b}}$ with $a \in \mathbb{Z}_N$ and $b \in [1, \mathrm{ord} QR_{N^2}^+]$, the Small Decisional Diffie-Hellman (S-DDH) problem is to differentiate $Z = \xi^{\underline{ab}}$ from an arbitrary group element in $QR_{N^2}^+$.

In [7], the authors prove the following results:

(i) The PDL problem in $QR_{N^2}^+$ is as hard as PDL problem in QR_{N^2}.
(ii) The PDL problem in $QR_{N^2}^+$ cannot be harder than factoring N.
(iii) The DDH problem in $QR_{N^2}^+$ is as hard as DDH problem in QR_{N^2}.
(iv) The DDH problem in $QR_{N^2}^+$ cannot be harder than factoring N.

4.4 Encryption Scheme

Hybrid ElGamal (HEG) Scheme

In a hybrid encryption scheme, encryption algorithm used a public key encryption scheme to encrypt a random symmetric key and then a message is encrypted using this symmetric key. The decryption algorithm used a public key scheme to recover the symmetric key, which is used to recover the message. The Hybrid ElGamal (HEG) scheme is a combination of a symmetric cipher and the ElGalmal scheme with a hash function for key derivation.

Hofheinz et al. [1] proposed a Hybrid ElGamal scheme over the group QR_N^+. They proved that their scheme is CCA secure under the SDH assumption. The scheme is defined as follows [1]:

Setup: Consider a symmetric cipher SE $= (E, D)$ with key-space $\mathcal{K} = \{0, 1\}^{\updownarrow(k)}$ and let $\mathcal{H} = (\mathcal{H}_k)_{k\in\mathbb{N}}$ denotes the set of hash functions with $H : \{0, 1\}^{2n(k)} \rightarrow \{0, 1\}^{\updownarrow(k)}$ for each $H \in \mathcal{H}_k$, $n(k)$ is a function. Also, $RSA_{gen'}(1^k)$ is an algorithm that computes elements (N, P, Q, S) such that:

(i) $N = PQ(n\text{-bits})$ is a Blum integer where prime factors of $\frac{\phi(N)}{4}$ are pairwise distinct and
(ii) $S > 1$ is a divisor of $\frac{\phi(N)}{4}$ with $1 < \gcd\left(S, \frac{P-1}{2}\right) < \frac{P-1}{2}$ and $1 < \gcd\left(S, \frac{Q-1}{2}\right) < \frac{Q-1}{2}$.

Key Generation

- Compute an RSA modulus $N = PQ$ using $RSA_{gen'}(1^k)$.
- Choose a generator ξ of QR_N^+.
- Choose an exponent $a \in \left\{1, \cdots, \frac{N}{4}\right\}$ and compute $X = \xi^{\underline{a}}$.
- Choose a hash function $H \in \mathcal{H}_k$.

The public key is (N, ξ, X, H) and the corresponding secret key is (N, a, H).

Encryption: Given a message$m$, the sender has to choose a random number $b \in \left\{1, \cdots, \frac{N}{4}\right\}$ and generate$Y = \xi^{\underline{b}}$, $K = H\left(Y, X^{\underline{b}}\right)$and $\psi = \mathrm{E_K}(m)$. The pair $(Y, \psi) \in QR_N^+ \times \{0, 1\}^*$ is the ciphertext of$m$.

Decryption: The receiver first verifies whether $Y \in QR_N^+$ or not and rejects if not. If $Y \in QR_N^+$, then compute $K = H(Y, Y^{\underline{a}})$ and recover the message by computing $D_K(\psi)$.

Twine Hybrid ElGamal (THEG) Scheme

Cui et al. [3] proposed a Twine Hybrid ElGamal (THEG) scheme over the group QR_N^+. They proved that their scheme is IND-CCA secure under Diffie-Hellman assumption. We refer to [3], for details security analysis of this scheme. The scheme is proposed as follows [3]:

Setup: Let SE $= (E, D)$ be a symmetric cipher with key-space $\mathcal{K} = \{0, 1\}^{\updownarrow(k)}$ and let $\mathcal{H} = (\mathcal{H}_k)_{k\in\mathbb{N}}$ be a set of hash functions with $H : \{0, 1\}^{3n(k)} \rightarrow \{0, 1\}^{\updownarrow(k)}$ for each $H \in \mathcal{H}_k$. Also, $RSA_{gen'}(1^k)$ is an algorithm which outputs the elements (N, P, Q, S) such that:

(i) $N = PQ$ (n-bits) is a Blum integer such that $\frac{\phi(N)}{4}$ has pairwise distinct prime factors and
(ii) $S > 1$ is a divisor of $\frac{\phi(N)}{4}$ with $1 < \gcd\left(S, \frac{P-1}{2}\right) < \frac{P-1}{2}$ and $1 < \gcd\left(S, \frac{Q-1}{2}\right) < \frac{Q-1}{2}$.

Key Generation:

- Generate an RSA modulus $N = PQ$ with $RSA_{gen'}(1^k)$.
- Choose a generator ξ of QR_N^+.

- Choose an exponent $a, b \in \{1, \cdots, \frac{N}{4}\}$ and compute $X = \xi^{a}$ & $Y = \xi^{b}$.
- Choose a hash function $H \in \mathcal{H}_k$.

The public key is (N, ξ, X, Y, H) and the corresponding secret key is (N, a, b, H).

Encryption: Given a message m, choose a random element $c \in \{1, \cdots, \frac{N}{4}\}$ and compute $Z = \xi^{c}$, $K = H\left(Z, X^{c}, Z^{c}\right)$ and $\psi = \mathrm{E_K}(m)$. The pair $(Z, \psi) \in QR_N^+ \times \{0, 1\}^*$ is the required ciphertext of$m$.

Decryption: The receiver first verifies whether $Z \in QR_N^+$ or not and rejects if not. If $Z \in QR_N^+$, then compute $K = H\left(Z, Z^{\underline{a}}, Z^{\underline{b}}\right)$ and recover the message by computing $D_K(\psi)$.

Key Encapsulation Mechanism (KEM)

A KEM consists of three algorithms: Key Generation, Encapsulation, and Decapsulation mechanism. For a security parameter k, a key generation algorithm provides a public key pk and private key sk. For pk, an encapsulation algorithm returns (K, C), where $K \in \{0, 1\}^{\updownarrow_K}$ is a symmetric key and C is a ciphertext. For the input (sk, C), decapsulation algorithm returns $K \in \{0, 1\}^{\updownarrow_K}$ or $\perp$. Note that an IND-CCA secure symmetric scheme along with IND-CCA secure KEM yield an encryption scheme having IND-CCA security [31]. Therefore, it is worthy to construct KEM having IND-CCA security. Because of indistinguishable property of signed quadratic residue, it is used to build KEM having IND-CCA security and hence a practical IND-CCA secure encryption scheme. Such applications are noticed in [4–6]. For the completeness of this section, we have considered the following schemes from [4–6]. We refer to [4–6] for detail description. Nevertheless, a target-collision resistant hash function T is a function, where for a random chosen pre-image ξ, it is difficult to find ξ' such that $T(\xi) = T(\xi')$. Here BBS (Blum-Blum-Shub) is a pseudorandom number generator and $\updownarrow_i$ refer for the length of i.

KEN Proposed in [4]

Key Generation:

- Choose a modulus $N = PQ = (2p + 1)(2q + 1)$.
- Choose a signed quadratic residue $\xi \in QR_N^+$.
- Choose exponent $a \in \{1, \cdots, \frac{N-1}{4}\}$.
- Compute $X = \xi^{a 2^{\updownarrow_K + \updownarrow_T}}$.

For a security parameter k, key generation algorithm returns a pair of keys (pk, sk), where $pk = (N, \xi, X)$ and $sk = (N, \xi, a)$.

Encapsulation:

- Choose a random element $b \in \{1, \cdots \frac{N-1}{4}\}$.
- Choose a target collision resistance hash function $T : QR_N^+ \to \{1, \cdots, 2^{\updownarrow_T} - 1\}$.
- Compute $R = \xi^{b 2^{\updownarrow_K + \updownarrow_T}}$, $t = T(R) \in \{1, \cdots, 2^{\updownarrow_T} - 1\}$, $S = \left(\xi^t X\right)^b$.

Encapsulation algorithm returns the key $K = \mathrm{BBS}_N(\xi^{b2^{\updownarrow_T}}) \in \{0, 1\}^{\updownarrow_K}$ and the ciphertext $C = (R, S) \in QR_N^+ \times QR_N^+$. Here, for $\xi \in QR_N^+$,

$$\mathrm{BBS}_N(\xi) \stackrel{\text{def}}{=} \left(\mathrm{LSB}_N(\xi), \mathrm{LSB}_N(\xi^2), \cdots, \mathrm{LSB}_N\left(\xi^{2^{\updownarrow_K - 1}}\right)\right) \in \{0, 1\}^{\updownarrow_K}$$

where $\mathrm{LSB}_N(\xi) = \xi \pmod 2$, is the last significant bit of ξ.

Decapsulation:

- Check $(R, S) \in QR_N^+ \times QR_N^+$ or not and reject if not.
- Compute $t = T(R) \in \{1, \cdots, 2^{\updownarrow_T} - 1\}$
- Check $S^{2^{\updownarrow_K + \updownarrow_r}} = R^{t + a2^{\updownarrow_K + \updownarrow_T}}$ or not and reject if not.
- Compute $\alpha, \beta, \gamma \in \mathbb{Z}$ such that $2^\gamma = \gcd(t, 2^{\updownarrow_K + \updownarrow_T}) = \alpha t + \beta 2^{\updownarrow_K + \updownarrow_T}$.
- Compute $T = (S^\alpha \cdot R^{\beta - a\alpha})^{2^{\updownarrow_T - c}}$.
- Compute $K = \mathrm{BBS}_N(T) \in \{0, 1\}^{\updownarrow_K}$, which is the output.

KEN Proposed in [5]

In [5], Mei et al. proposed the following CCA secure encryption algorithm.

Key Generation:

- Choose a modulus $N = PQ = (2pp' + 1)(2qq' + 1)$, where p', q' are primes and p, q are both product of distinct odd primes smaller than a lower bound B.
- Choose a target collision resistance hash function $T : \mathbb{Z}_N \to \{1, \cdots, 2^{\updownarrow_H} - 1\}$.
- Choose a random element $\xi \in G$, a bit string r of length $\updownarrow_{(N-1)/2}$, and ρ, ρ' from $\left\{1, \cdots, 2^{\updownarrow_{p'} + \updownarrow_{q'} + \lambda}\right\}$.
- Compute $X = \xi^{\rho 2^v} (v = \updownarrow_T + \updownarrow_K)$ and $X' = \xi^{\rho'}$

The Key generation algorithm returns (pk, sk). Here $pk = (N, \xi, X, X', r, H)$ and $sk = (\rho, \rho')$.

Encapsulation:

- Choose $\mu \in \left\{1, \cdots, 2^{\updownarrow_{p'} + \updownarrow_{q'} + \lambda}\right\}$.
- Computes $R = \left|\xi^{\mu 2^v}\right|, t = T(R), S = \left|\left(X'^t X\right)^\mu\right|, H = |X'^{\mu 2^v}|$

Algorithm returns the key $K = \mathrm{BBS}_r^+(H)$ and the ciphertext $C = (R, S)$. Here

$$\mathrm{BBS}_r^+(H) \stackrel{\text{def}}{=} \left(B_r(|H|), B_r(|H^2|), \cdots, B_r\left(\left|H^{2^{\updownarrow_K - 1}}\right|\right)\right)$$

Decapsulation:

- Check $(R, S) \in QR_N^+ \times QR_N^+$ or not and reject if not. Otherwise go to next step.
- Compute $t = T(R) \in \{1, \cdots, 2^{\updownarrow_T} - 1\}$.
- Check $|S^{2^v}| = \left|R^{\rho' t + \rho 2^v}\right|$ or not and reject if not. Otherwise go to next step.

- Compute $H = \left| R^{\rho'} \right|$.
- Compute $K = \mathrm{BBS}_r^+(H)$, which is the output.

KEN Proposed in [6]

In 2013 [6], Qin et al. proposed a KEM defined over QR_N^+, which is the extension KEM proposed in [4].

Key Generation:

- Choose a modulus $N = PQ = (2p+1)(2q+1)$.
- Choose a signed quadratic residue $\xi \in QR_N^+$.
- Choose exponent $a, b \in \left\{1, \cdots, \frac{N-1}{4}\right\}$.
- Compute $X = \xi^{1-a2\updownarrow_K+\updownarrow_T}$ and $Y = \xi^{-b2\updownarrow_K+\updownarrow_T}$.

The key generation algorithm returns (pk, sk), where $pk = (N, \xi, X, Y)$ and a $sk = (N, \xi, a, b)$.

Encapsulation:

- Choose a random element $r \in \left\{1, \cdots \frac{N-1}{4}\right\}$.
- Choose a target collision resistance hash function $T : \mathbb{Z}_N \to \{1, \cdots, 2^{\updownarrow_T} - 1\}$.
- Compute $R = \xi^{r2\updownarrow_K+\updownarrow_T}$, $t = T(R)$ and $S = \left(XY^t\right)^r$.

Encapsulation algorithm returns the key $K = \mathrm{BBS}_N\left(\xi^{r2\updownarrow_T}\right) \in \{0,1\}^{\updownarrow_K}$ and the ciphertext $C = (R, S) \in QR_N^+ \times QR_N^+$.

Decapsulation:

- Check $(R, S) \in QR_N^+ \times QR_N^+$ or not and reject if not.
- Compute $t = T(R) \in \{1, \cdots, 2^{\updownarrow_T} - 1\}$.
- Check $S^{2\updownarrow_K+\updownarrow_r} = R^{1-(a+tb)2\updownarrow_K+\updownarrow_T}$ or not and reject if not.
- Compute $T = (SR^{a-tb})^{2\updownarrow_T}$
- Compute $K = \mathrm{BBS}_N(T) \in \{0,1\}^{\updownarrow_K}$, which is the output.

PKE with double decryption mechanism (DD-PKE Scheme)

At Eurocrypt'99, Bresson et al. [24] showed a different approach over the group QR_{N^2}, where N is a Blum integer. Their scheme (called BCP cryptosystem) has two decryption procedures, namely, user decryption and master decryption procedure. Because of two decryption procedures, it has many applications in real-life scenarios. The public key cryptosystem having two decryption procedures is referred to as DD-PKE scheme. The idea of DD-PKE scheme is first introduced by Cramer and Shoup [32]. A DD-PKE scheme consists of four algorithms: Key generation, Encryption, User Decryption, and Master Decryption. For statutory definition of DD-PKE we refer to [33]. In 2019, Goswami et al. [7] pointed out an attack on BCP cryptosystem and proposed a modified BCP cryptosystem over $QR_{N^2}^+$, where N is a Blum integer. Their scheme is based on PDL problem over $QR_{N^2}^+$. Their scheme is one-way iff the S-CDH assumption holds in $QR_{N^2}^+$ and the semantic security of the scheme is based on the S-DDH assumption over $QR^+_{N^2}$. For details we refer to [7]. The modified BCP cryptosystem is proposed as follows [7]:

Setup: For a security parameter k, compute the following:

- Choose k bit length $N = PQ$, where P and Q are safe primes.
- Choose a generator $\xi \in QR^+_{N^2}$.

Here, $\mathbb{Z}_N$ is the message space. The triplet (k, N, ξ) is considered as public parameter PK and the master secret key SK is (P, Q).

Key Generation: Choose a random element $a \in \left[1, \text{ord}\, QR^+_{N^2}\right]$ and compute $\mu = \xi^{\underline{a}} = \left|\xi^a (\text{mod} N^2)\right|$. The public key is (N, ξ, μ) and secret key is a.

Encryption: For a given message $m \in \mathbb{Z}_N$, choose an element $r \in \mathbb{Z}_N$ and compute $A = \xi^{\underline{r}} = \left|\xi^r \pmod{N^2}\right|$ and $B = \mu^{\underline{r}}|1 + mN| = \left|\mu^r(1 + mN) \pmod{N^2}\right|$. The ciphertext of m is (A, B).

User Decryption: A user having secret key a can recover the message m as follows: compute

$A^{\underline{a}} = \xi^{\underline{ar}} = |\xi^{ar} \pmod{N}^2|$ and

$$\frac{B}{A^{\underline{a}}} = \left|\frac{B}{A^a} \pmod{N^2}\right| = \left|(1 + mN) \pmod{N^2}\right|$$

Thus, $m = \frac{\frac{B}{A^{\underline{a}}} - 1 \pmod{N^2}}{N}$ or $\frac{-\frac{B}{A^{\underline{a}}} - 1 \pmod{N^2}}{N}$. Since either$N \mid \left(\frac{B}{A^{\underline{a}}} - 1\right)orN \mid \left(-\frac{B}{A^{\underline{a}}} - 1\right)$, so user can recover the original message$m$.

Master Decryption: For a user who knows the factors of N can compute $r \pmod{N}$ as follows:

$$r = \frac{A^{\underline{\lambda(N)}} - 1 \pmod{N^2}}{N} \cdot k^{-1} \pmod{N}$$

or

$$r = \frac{-A^{\underline{\lambda(N)}} - 1 \pmod{N^2}}{N} \cdot k^{-1} \pmod{N}$$

Therefore m can be determined by $\frac{\frac{B}{\mu^{\underline{r}}} - 1 \pmod{N^2}}{N}$ or $\frac{-\frac{B}{\mu^{\underline{r}}} - 1 \pmod{N^2}}{N}$.

Self-bilinear map

Bilinear maps are important tools for constructing many cryptographic primitives, such as attribute-based encryption (ABE), identity-based encryption (IBE), and non-interactive zero-knowledge (NIZK) proof systems. However, most of the bilinear maps which are used in cryptography are defined over the elliptic curve groups. A bilinear map is called *self-bilinear map* if the domain and target groups are equivalent.

It is studied by Cheon and Lee [34]. They showed that a self-bilinear map can be used for constructing cryptographic primitives such as in multilinear map.

Let G be a cyclic group. A map $e : G \times G \rightarrow G$ is called a self-bilinear map if

- $e(\xi_1^a, \xi_2) = e(\xi_1, \xi_2^a) = e(\xi_1, \xi_2)^a, \forall \xi_1, \xi_2 \in G$ and $a \in \mathbb{Z}$
- If $\xi_1, \xi_2 \in G$ are generators of G, then $e(\xi_1, \xi_2)$ is a generator of G, i.e., the map e is non-degenerate.

Note that, for any integer $n \geq 2$, a n-multilinear map e_n can be constructed from a self-bilinear map e. The n-multilinear map e_n is defined as

$$e_n(\xi_1, \ldots, \xi_{n-1}, \xi_n) := e(e_{n-1}(\xi_1, \ldots, \xi_{n-1}), \xi_n).$$

In [8], Yamakawa et al. introduced a weaker variant of self-bilinear map called *self-bilinear map with auxiliary information*. They showed that in spite of the fact that it is a weaker version still it proves to be useful to build some cryptographic primitives. They have considered the group QR_N^+, where N is a Blum integer. They define a self-bilinear map as follows:

$$e : QR_N^+ \times QR_N^+ \rightarrow QR_N^+$$

$$e(\xi_a, \xi_b) := \xi^{2ab}$$

It is observed that, even if one can compute $e(\xi^a, \xi^b) = \xi^{2ab}$, it is difficult to compute ξ^{ab} from it as computing square root modulo N is as hard as finding the factors of N. Again, it is difficult to compute $e(\xi^a, \xi^b)$ for given ξ^a, ξ^b, without any auxiliary information. They defined an auxiliary information τ_b for each element $\xi^b \in QR_N^+$ which helps to compute $e(\cdot, \xi^b)$ efficiently. They define τ_b as a circuit which can compute the t_b-th power, where $t_b = 2b \pm \text{ord}(QR_N^+)$. But without the knowledge of the factorization of N there is no efficient algorithm to compute t_b from b. To overcome this difficulty, author used an efficient randomized algorithm called indistinguishability obfuscation ($i\mathcal{O}$) that makes circuits C_0 and C_1 identical when they function exactly in the same way. They observed that a circuit C_0 that computes the $2b$-th power and a circuit C_1 that computes the t_bth power for elements of QR_N^+ are identical since $t_b = 2b \pm \text{ord}(QR_N^+)$. Therefore, one will be obfuscated by the circuit C_0 that computes the $2b$-th power for elements of QR_N^+. For details we refer to [8].

Pseudo-free group

The concept of *pseudo-free group* was initially introduced by Hohenberger [35] in the year 2003, however it was later strengthened by Rivest [36]. Apparently, a pseudo-free group differs very little from a free group. In fact, a pseudo-free group G behaves like a free group but in case of a pseudo-free group G, a PPT adversary can neither generate an equation E nor a solution S efficiently for E in G, if E has no solution in the corresponding free group. A more precise definition of pseudo-free group can be found in [36]. However, to assume a group to be a pseudo-free group is rather a strong assumption. In his paper [36], Rivest conjectured that for a Blum integer N, the group $\mathbb{Z}_N^*$ is a pseudo-free group and later Micciancio [37], proved that $\mathbb{Z}_N^*$ is a pseudo-free group under strong RSA assumption by choosing the elements from QR_N randomly. He asked whether the proof can be carried over if sample elements are chosen at random from the whole group $\mathbb{Z}_N^*$. In 2009 [9], Jhanwar et al. proved that RSA group is pseudo-free group if the elements are chosen randomly from QR_N^+ (Strong SQR-RSA problem). They also mentioned that RSA group is pseudo-fee group if the elements are chosen randomly from $QR_N \cup QR_N^+$. For more details we refer to [9].

Static TDDH Group

In 2006 [38], Dent-Galbraith introduced the concept of *Trapdoor DDH* (TDDH) group, a group where DDH problem is hard to solve without any additional information. It is used to build up identification scheme [38] and statistically hiding set [39]. In [10], Seurin introduced a variant of TDDH groups and called *Static TDDH* (STDDH) group. A STDDH group is similar to a TDDH group, where the trapdoor is such that for a fixed pair of group element (ξ, ξ^a), one can solve DDH problem. In [10], authors used signed quadratic residue to construct Static TDDH group (called $\mathcal{STDDH}_{\text{SQR}}$). They proved that $\mathcal{STDDH}_{\text{SQR}}$ is STDDH group under the factoring assumption and the DDH assumption for QR_N^+. For details we refer to [10].

5 Conclusion

In this paper, we discussed the properties and behavior of SQR and presented some of its applications to cryptography. Some other cryptographic applications of SQR can be found in [12–15]. Despite their various interesting properties, they have found a little cryptographic application so far. To bring out some of these applications together was the main motivation behind this paper.

References

1. Hofheinz D, Kiltz E (2009) The group of signed quadratic residues and applications. In: Annual international cryptology conference. Springer, Berlin pp 637–653
2. Okamoto T, Pointcheval D (2001) The gap-problems: a new class of problems for the security of cryptographic schemes. In: International workshop on public key cryptography. Springer, Berlin, pp 104–118
3. Cui Y, Liu Y (2010) Twin hybrid ElGamal encryption over signed quadratic residue groups. In: International conference on wireless algorithms, systems, and applications. Springer, Berlin, pp 282–286
4. Hofheinz D, Kiltz E, Shoup V (2013) Practical chosen ciphertext secure encryption from factoring. J Cryptol 26(1):102–108

5. Mei Q, Li B, Lu X, Jia D (2011) Chosen ciphertext secure encryption under factoring assumption revisited. In: International workshop on public key cryptography. Springer, Berlin, pp 210–227
6. Qin B, Liu S (2013) Efficient chosen ciphertext secure public-key encryption under factoring assumption. Secur Commun Netw 6(3):351–360
7. Goswami P, Singh MM (2018) An application of signed quadratic residues in public key cryptography. Discret Math, Algorithms Appl 10(6):1850081
8. Yamakawa T, Yamada S, Hanaoka G, Kunihiro N (2017) Self-bilinear map on unknown order groups from indistinguishability obfuscation and its applications. Algorithmica 79(4):1286–1317
9. Jhanwar MP, Barua R (2009) Sampling from signed quadratic residues: RSA group is pseudofree. In: International conference on cryptology. Springer, Berlin, pp 233–247
10. Seurin Y (2013) New constructions and applications of trapdoor DDH groups. In: International workshop on public key cryptography. Springer, Berlin, pp 443–460
11. Zhang J, Wang XA (2013) New construction of PVPKE scheme based on signed quadratic residues. In: 5th international conference on intelligent networking and collaborative systems. IEEE, pp 434–437
12. Liu Z, Yang X, Zhong W, Han Y (2014) An efficient and practical public key cryptosystem with CCA-security on standard model. Tsinghua Sci Technol 19(5):486–495
13. Zhang M, Wang XA, Yang X, Li W (2015) New construction of PVPKE scheme and its application in information systems and mobile communication. Mobile Inf Syst. https://doi.org/10.1155/2015/430797
14. Wee H (2012) Public key encryption against related key attacks. In: International workshop on public key cryptography. Springer, Berlin, pp 262–279
15. Stinson DR, Paterson M (2018) Cryptography: theory and practice, 4th edn. CRC Press
16. Rivest RL, Shamir A, Adleman L (1978) A method for obtaining digital signatures and public-key cryptosystems. Commun ACM 21(2):120–126
17. Goldwasser S, Micali S (1984) Probabilistic encryption. J Comput Syst Sci 28(2):270–299
18. Lei F, Chen W, Chen K (2006) A non-committing encryption scheme based on quadratic residue. In: International symposium on computer and information sciences. Springer, Berlin, pp. 972–980
19. Harn L, Kiesler T (1996) An efficient probabilistic encryption scheme. Inf Process Lett 34(3):123–129
20. Cocks C (2001) An identity-based encryption scheme based on quadratic residues. In: IMA international conference on cryptography and coding. Springer, Berlin, pp 360–363
21. Rabin MO (1979) Digitalized signatures and public-key functions as intractable as factorization. Massachusetts Inst. of Tech., Cambridge Lab for Computer Science
22. Williams H (1980) A modification of the RSA public-key encryption procedure (Corresp.). IEEE Trans Inf Theory. 26(6):726–729
23. Galindo D, Martýn S, Morillo P, Villar JL (2003) A practical public key cryptosystem from Paillier and Rabin schemes. In: International workshop on public key cryptography. Springer, Berlin, pp 279–291
24. Bresson, E., Catalano, D., Pointcheval, D.: A simple public-key cryptography with a double trapdoor decryption mechanism and its applications. In: Proceedings of advances in cryptology—ASIACRYPT 2003, pp 37–54
25. Goswami P, Singh MM, Bhuyan B (2017) A new public key cryptosystem over Zn2*. Discret Math, Algorithms Appl 9(6):1750080
26. Shmuely Z (1985) Composite Diffie-Hellman public-key generating systems are hard to break. Technical Report no. 356
27. McCurley KS (1988) A key distribution system equivalent to factoring. J Cryptol 1(2):95–105
28. Barić N, Pfitzmann B (1997) Collision-free accumulators and fail-stop signature schemes without trees. In: International conference on the theory and applications of cryptographic techniques. Springer, Berlin, pp 480–494
29. Cramer R, Shoup V (2000) Signature schemes based on the strong RSA assumption. ACM Trans Inf Syst Secur (TISSEC) 3(3):161–185

30. Paillier P (1999) Public-key cryptosystems based on composite degree residuosity classes. In International conference on the theory and applications of cryptographic techniques. Springer, Berlin, pp 223–238
31. Cramer R, Shoup V (2003) Design and analysis of practical public-key encryption schemes secure against adaptive chosen ciphertext attack. SIAM J Comput 33(1):167–226
32. Cramer R, Shoup V (2002) Universal hash proofs and a paradigm for adaptive chosen ciphertext secure public-key encryption. In: International conference on the theory and applications of cryptographic techniques. Springer, Berlin, pp 45–64
33. Kevin H (2008) The theory and applications of Homomorphic cryptography. University of Waterloo, Canada, Computer Science
34. Cheon JH, Lee DH (2009) A note on self-bilinear maps. Bull Korean Math Soc 46(2):303–309
35. Hohenberger SR (2003) The cryptographic impact of groups with infeasible inversion (Doctoral dissertation, Massachusetts Institute of Technology)
36. Rivest RL (2004) On the notion of pseudo-free groups. In: Theory of cryptography conference. Springer, Berlin, pp 505–521
37. Micciancio D (2005) The RSA group is pseudo-free. In: Annual international conference on the theory and applications of cryptographic techniques. Springer, Berlin, pp 387–403
38. Dent AW, Galbraith SD (2006) Hidden pairings and trapdoor DDH groups. In: International Algorithmic number theory symposium. Springer, Berlin, pp 436–451
39. Prabhakaran M, Xue R (2009) Statistically hiding sets. In: Cryptographers' track at the RSA conference. Springer, Berlin, pp 100–116

The manufacturer's authorised representative in the EU is Springer Nature Customer Service Centre GmbH, Europaplatz 3, 69115 Heidelberg, Germany. If you have any concerns regarding our products, please contact ProductSafety@springernature.com

Printed and bound by CPI Group (UK) Ltd, Croydon, CR0 4YY
01/10/2026
02225855-0005